AF833616

TELECOM SECTOR REGULATION IN INDIA

This book traces important legal and regulatory developments in the first two decades since the Telecom Regulatory Authority of India (TRAI) was established, along with its political and economic aspects. It narrates the story of the institutional progress of TRAI and its influence on the growth of India's telecom sector.

The telecom revolution was a game changer in post-liberalization India, a country today home to the second largest subscriber base in the world– more people have access to mobile phones than toilets. Its rapid, relentless growth has created new possibilities and challenges, including a robust regulatory policy. This book, the first comprehensive survey of TRAI's progress, examines the salient developments in regulation of the Indian telecom sector. It analyses, at the macro-institutional level, the norms and rules reconstituted over time; at the institutional level, the impact of important court judgments, relevant telecom case law (including the 2G judgment and Adjusted Gross Revenue-related cases), and the 'judicialization' of regulatory governance; and, at the micro-institutional level, the mechanisms of governance of TRAI and the way its functioning has affected the alignment of incentives in the regulatory space. It provides an overview of the regulatory framework and the context in which the telecom sector was deregulated, the structure of internal governance, and issues in telecom licensing and spectrum allotment.

The book combines academic rigour and empirical research with a practitioner's perspective of the unfolding events. It will interest students and researchers of economics, law, public policy, communications technology, and ICT policy and regulation, as well as telecom sector professionals, service providers, academic experts, policymakers, and think tanks.

Maruthi P. Tangirala is Deputy Director General, Department of Telecommunications, Ministry of Communications, New Delhi, India, and an officer of the Indian P&T Accounts and Finance Service (1990 batch). He holds advanced degrees in Management (Indian Institute of Management, Calcutta), Public Administration

(Panjab University/The Indian Institute of Public Administration), and Law and Governance (Jawaharlal Nehru University), in addition to degrees in Civil Engineering (College of Engineering, Guindy) and Law (Osmania University). He has worked in the Government of India's Department of Telecommunications and in Telecom Regulatory Authority of India, Union Public Service Commission, and Insurance Regulatory and Development Authority of India, as well as in Bharat Sanchar Nigam Limited, in different functional areas. He has published scholarly articles and opinion pieces in the *Economic and Political Weekly*, *Competition and Regulation in Network Industries*, and *Indian Express*, among others.

TELECOM SECTOR REGULATION IN INDIA

An Institutional Perspective

Maruthi P. Tangirala

Routledge
Taylor & Francis Group
LONDON AND NEW YORK

First published 2019 by Routledge

2 Park Square, Milton Park, Abingdon, Oxon, OX14 4RN
605 Third Avenue, New York, NY 10017

Routledge is an imprint of the Taylor & Francis Group, an informa business

First issued in paperback 2020

Copyright © 2019 Maruthi P. Tangirala

The right of Maruthi P. Tangirala to be identified as author of this work has been asserted by him in accordance with sections 77 and 78 of the Copyright, Designs and Patents Act 1988.

All rights reserved. No part of this book may be reprinted or reproduced or utilised in any form or by any electronic, mechanical, or other means, now known or hereafter invented, including photocopying and recording, or in any information storage or retrieval system, without permission in writing from the publishers.

Notice:
Product or corporate names may be trademarks or registered trademarks, and are used only for identification and explanation without intent to infringe.

British Library Cataloguing-in-Publication Data
A catalogue record for this book is available from the British Library

Library of Congress Cataloging-in-Publication Data
Names: Tangirala, Maruthi P., author.
Title: Telecom sector regulation in India : an institutional perspective / Maruthi P. Tangirala.
Description: Abingdon, Oxon ; New York, NY : Routledge, 2019. | Includes bibliographical references and index.
Subjects: LCSH: Telecommunication—India. | Telecommunication policy—India.
Classification: LCC HE8374 .T36 2019 | DDC 384.0954—dc23
LC record available at https://lccn.loc.gov/2018057022

ISBN: 978-1-138-57291-1 (hbk)
ISBN: 978-0-367-73143-4 (pbk)

Typeset in Sabon
by Apex CoVantage, LLC

TO MY LATE PARENTS, AFFECTIONATELY

CONTENTS

ILLUSTRATIONS

Figure

Tables

ABBREVIATIONS

2G	second generation (mobile technology)
3G	third generation (mobile technology)
ABTO	Association of Basic Telecom Operators
ADC	access deficit charge
AGR	adjusted gross revenue
ASR	accounting separation regulation
BICP	Bureau of Industrial Costs and Prices
BITs	bilateral investment treaties
BSNL	Bharat Sanchar Nigam Limited
BSO	basic service operator
BWA	broadband wireless access
CAA	Coastal Aquaculture Authority
CAC	command and control
CAG	comptroller and auditor general
CCI	Competition Commission of India
CDMA	Code Division Multiple Access
CERC	Central Electricity Regulatory Commission
CMSP	cellular mobile service provider
CMTS	cellular mobile telephone service
COAI	Cellular Operators Association of India
CPC	Code of Civil Procedure, 1908
CPP	calling party pays
CrPC	Code of Criminal Procedure, 1973
DoT	Department of Telecommunications (government of India)
EM	explanatory memorandum
EV-DO	enhanced voice-data optimized
FCFS	first come first served
FDI	foreign direct investment
FMC	Forward Markets Commission
FSP	fixed service provider

FWT	fixed wireless terminal
GMPCS	global mobile personal communications service
GoM	group of ministers
GoT	group on telecom
GoT-IT	group on telecom and information technology
GSM	Group Speciale Mobile
ICICI	Industrial Credit and Investment Corporation of India Limited
ICT	information and communication technology
ILD	international long distance
ILDO	international long distance operator
IMT-2000	International Mobile Telecommunications 2000
IPC	Indian Penal Code
IRA	independent regulatory authority
IRDAI	Insurance Regulatory and Development Authority of India
ISPs	internet service providers
IT	information technology
ITA	Indian Telegraph Act, 1885
ITU	International Telecommunications Union
IUC	interconnection usage charges
JPC	Joint Parliamentary Committee
LoI	letter of intent
M&A	mergers and acquisitions
MIB	Ministry of Information and Broadcasting
MNP	mobile number portability
MoD	Ministry of Defence
MoF	Ministry of Finance
MRTPC	Monopolies and Restrictive Trade Practices Commission
MTC	mobile termination charges
MTNL	Mahanagar Telephone Nigam Limited
NCDRC	National Consumer Disputes Redressal Commission
NEP	new economic policy
NGO	non-governmental organization
NIA	Notice Inviting Applications
NIE	new institutional economics
NLD	national long distance
NLDO	national long distance operator
NPV	net present value
NTP 1994	National Telecom Policy 1994
NTP 1999	New Telecom Policy 1999
NTP 2012	National Telecom Policy 2012
OSP	other service provider

PAC	Public Accounts Committee
PMRTS	public mobile radio trunking service
PNGRB	Petroleum and Natural Gas Regulatory Board
PSTN	public switched telephone network
PTM	part time member
QoS	quality of service
RBI	Reserve Bank of India
SC	Supreme Court
SCC	Standing Committee on Communications
SCIT	Standing Committee on Information Technology
SDCA	short distance charging area
SEBI	Securities and Exchange Board of India
SLC	subscriber linked criteria
SUC	spectrum usage charges
TDR	triadic dispute resolution
TDSAT	Telecom Disputes Settlement and Appellate Tribunal
TEC	Telecommunications Engineering Centre
TIR 1999	Telecommunications Interconnection (Charges and Revenue Sharing) Regulation, 1999
TRAI	Telecom Regulatory Authority of India
TTO 1999	Telecommunications Tariff Order, 1999
UAS	unified access service
UASL	unified access service licence
UL	unified license
USL	universal service levy
USO	universal service obligation
VAS	value-added services
VSAT	very small aperture terminal
VSNL	Videsh Sanchar Nigam Limited
WLL	wireless in local loop
WPC	Wireless Planning and Coordination wing (DoT)
WTM	whole time member
WTO	World Trade Organization

ACKNOWLEDGEMENTS

I thank the government of India for granting me study leave for two years to pursue a PhD at the Centre for the Study of Law and Governance, Jawaharlal Nehru University – much of the research for this book was conducted during this period. Views expressed are personal and do not implicate the government of India in any way.

My colleagues and friends in the Department of Telecommunications, in TRAI, in the telecom sector, and in academia who have helped me with policy and regulatory insights are too numerous to name. Among them are Jaivir Singh, Niraja Gopal Jayal, Amita Singh, Amit Prakash, Pratiksha Baxi, Rahul Khullar, R. Chandrashekhar, Vijayalakshmy Gupta, Sadhana Dikshit, S. R. Rao, Annie Moraes, R. K. Arnold, G. Alagarsamy, Anuradha Mitra, Manish Sinha, Shahbaz Ali, Shahnawaz Alam, Sanjay Kumar, N. Parameswaran, Sudhir Gupta, K. Ramchand, Agneshwar Sen, Amit Govil, C.P.S. Bakshi, Robert Ravi, Arvind Kumar, Sunil Singhal, R. K. Gupta, Akhilesh Trivedi, Rohit Khare, Rajat Kathuria, D. Krishnan, Ashish Narayan, Piyush Choudhary, and Anupam Varma. Among those who helped in other ways were P.R.K. Rao, C. Shambu Prasad, Amitabh Kharkwal, Saurabh Kumar, and Prakash Rani Chawla. I thank the commissioning and editorial team at Routledge India for their support.

I am grateful to all my family members, specially Rashmi, Krishna, and Kavya, for bearing with me throughout.

Part I

OVERVIEW

1

INTRODUCTION

1.1 Background

The passage of about two decades, marked by rapid change in the business landscape and in administrative arrangements in the Indian telecommunications services sector, offers an opportunity to study significant aspects of the telecom sector's regulatory governance[1] from an institutional perspective. Apart from such an in-period assessment, the way telecom regulation has evolved and become structured in India also encourages the reflexive aspiration to examine some foundational assumptions constituting the governance of the sector. The present book is thus shaped not merely by the efflux of time but more crucially by fundamental constitutional questions concerning regulation and governance in the telecom sector.

Deregulation of the telecom sector

The World Trade Organization (WTO)'s Basic Agreement on Telecommunications in 1997 is recognized as the "definitive moment" in the transition of the telecom sector to a competitive marketplace (Walden and Angel 2007: 4).[2] The conclusion of extended negotiations in WTO that began in May 1994 and ended in February 1997 resulted in commitments to liberalize the telecom sector by a large number of countries, and the markets represented by the commitments accounted for more than 91 per cent of global telecom revenues (Tuthill and Sherman 2008). The propensity to commit at the conclusion of the negotiations was seen as evidence of the crumbling of conventional wisdom about the natural monopoly features of telecom infrastructure and the new role of telecom regulation as managing the transition from monopoly to market over a long-term time horizon (Tuthill 1997), which was reflected in a set of principles covering topics such as competition safeguards, interconnection guarantees, transparent licensing processes, and the independence of national regulators (Tuthill and Sherman 2008). Almost in parallel and as if in preparation, the Indian attempts at moving away from monopoly production and provision of telecommunications services began in the early 1990s, were tentatively articulated in the National Telecom Policy 1994 (NTP 1994),[3] attained critical mass with the enactment of the Telecom Regulatory Authority of India (TRAI) Act in 1997, and culminated in the formulation of the New Telecom Policy 1999 (NTP 1999), which managed to serve the prevailing policy consensus until executive overreach occasioned by inherent institutional imbalances led to a course correction.[4] Early moves to redraw the Indian telecom regulatory landscape also paralleled recognition – at the level of other international forums, such as the International Telecommunications Union (ITU) – of the changing nature of the telecom environment, where the private sector was coming to be seen as playing the lead role. Also recognized was a corresponding need to enhance the ITU's character as a partnership between the public and private sectors, making it the fundamental strategic premise of that organization (Codding 1995).

The nature of technology in telecom markets around the world altering the focus of telecom regulation is recognized in the literature (Tuthill 1997; Renda 2012). The transition from a market dominated by delivery of service through landline telephones to one where handheld wireless devices have become phones of first choice has coincided with the rise of the new regulatory architecture in India, with the architecture itself being shaped by the perceived requirement to attract private capital to both widen and deepen service coverage. Indian telecom sector regulation has evolved along its own unique 'hybrid' trajectory that simultaneously adapted elements from different regulatory theories and models, at times optimally and at other times suboptimally. The story of this transition is punctuated by struggles over policy, much litigation, and a generous amount of self-congratulation among policy actors and other stakeholders about the correctness of policy strategies and the achievement of policy 'targets'. The institutional interplay between the political and economic subsystems on the one hand and the telecom regulatory-administrative domain on the other – and the patterns that emerge from that interplay – is the focus of this book.

It bears noting that telecommunications services are characterized by features that tend to uniquely affect their regulation. The telecom services sector was seen throughout its history as a natural monopoly; in the early days of the development of telecom, most governments around the world viewed the newly emerging telecom services "primarily as government social services, not services to be supplied by private businesses in normal markets" (Melody 1997a: 11). Voice telecom (along with electricity) has been considered an industry closely bound up with the nation-state, and "the rise of big business, the welfare state, the mixed economy, and the affluent society of the post-war era are all mirrored in [its] development, consolidation, and governance" (Levi-Faur 1999: 176). Telephone services have been seen as public utility services that are required to deal reasonably with the public.

Telecom shares three important characteristics with other utility services: (1) economies of scale and scope, (2) asset specificity and non-redeployability (sunk costs), and (3) "a broad range of domestic users, usually overlapping the voting population of the country" (Levy and Spiller 1994: 203). It has traditionally been regarded as "a relatively straightforward public utility", where "economies of scale, political and military sensitivities, and large externalities made telecommunications a typical public service believed to be a natural monopoly" (Wellenius and Stern 1994: 2). Public utility regulators "have a variety of different tasks which cannot be reduced to any single logic, economic or otherwise" (Prosser 1999: 199), but the importance of economic analysis is evident even within a broadly pluralist conception of telecom regulation. Telecom is suitable for application of the four principal components of economic regulation that in combination distinguish public utility from other sectors of the economy: "control of entry, price fixing, prescription of quality and conditions of service, and the imposition of an obligation to serve all applicants under reasonable conditions" (Kahn 1970: 3). The purely engineering orientation of telecom service provision came to be transformed as realization dawned in the 1960s and 1970s that the service could be harnessed as a powerful tool of economic and social development, given the rise of information "as a fundamental factor of production, alongside capital and labor" (ibid). The ITU's Maitland Report, for instance, recognized that the existence of "an efficient telecommunications system confers direct and indirect benefits which entitle it to be regarded as a public good" (ITU 1984: 8). Walden and Angel list the important economic features of the telecom industry: (1) its multi-product nature; (2) the non-storability of its services; (3) stochastic (random) time-varying demands; (4) capacity constraints and sunk costs; (5) the existence of externalities between users; (6) natural monopoly elements; and (7) the complex vertical structure of the industry. The first five influence tariff-setting procedures, the last three constitute market failures, and the very last feature has implications for "the ease with which regulation is carried out" (2007: 31).

Developments in the Indian policy regime

The growth of the telecommunications services sector in India in the past two decades is considered an unprecedented success story. Although India adopted second generation digital technology fairly late, it has the second largest subscriber base in the world today (Sridhar and Prasad 2011) and almost one billion of the estimated world total of eight billion telephone subscriptions (TRAI 2015b). The primary policy objective of the early years (NTP 1994, 1999) of expanding telecommunications access has been fulfilled to a large extent in the sector, which has seen a transformation from government monopoly to competitive service provision and private sector ascendancy. The relevance of telephony growth to overall economic growth in developed and developing countries has been well documented (A. Hardy 1980; Saunders et al. 1983; Cronin et al. 1991; Röller and Waverman 2001; Sridhar and Sridhar 2004; Waverman et al. 2005; Chakraborty and Nandi 2011), as has the impact of the political economy and the institutional environment on the pace, sequence, and scope of telecommunication restructurings (J. P. Singh 1998, 2000), and considerable satisfaction has been expressed for the manner in which India has been able to accelerate telephony growth (Kathuria et al. 2009). The development impact of mobile telephony in India has also been documented and commented upon (Tangirala 2009). However, the Indian telecom sector has also witnessed considerable controversy and policy anxiety from time to time in the post-liberalization phase since the early 1990s. The formulation of NTP 1999 within five years of NTP 1994 is an example, as is the sudden switch to NTP 2012 in the wake of the adverse judicial attention the telecom sector received around the time the policy was conceived.[5]

Thus, the policy and regulatory regime for the sector has undergone significant changes in the past two decades. It is not often acknowledged that India was pushing for telecom reforms – albeit "halting and nepotistic" (J. P. Singh 1998: 55) – as early as the mid-1980s, when such reform initiatives were rare in the developing world (Petrazzini 1996). However, there was little progress, and the regime's essential features remained the same well into the mid-1990s. The government was the monopoly provider of most telecom services and also played the role of policymaker and regulator in the earlier era. The enactment of the Telecom Regulatory Authority of India Act, 1997 (TRAI Act), which created TRAI as an independent regulatory authority (IRA), was a milestone. The Department of Telecommunications (DoT) continued to be the policymaker, retaining its role in licensing, spectrum provisioning, and standards setting. After some experience, several court cases, and based further on the prevailing policy vision, the TRAI Act was amended in 2000 to separate regulatory and adjudicatory functions by creating a Telecom Disputes Settlement and Appellate Tribunal (TDSAT) in addition to TRAI.[6] TRAI has been an active regulator and policy advisor and has taken credit for the telecom 'success story' attributed to a three-way partnership between government, regulator, and the private sector (TRAI 2007d).

The success story has been somewhat attenuated recently with questions being raised in different forums about perceived favouritism and violation of rules in policy formulation and implementation at key moments during the past two decades. Regulatory experience in India has seen the evolution, resolution, and recalcitrance of disputes among key actors in the regulatory arena. These disputes have not been characterized by any degree of consistency of position of the disputants or by any particular clarity in the articulation of the underlying motivations. While some were amenable to legislative and executive resolution, quite a few others have had judicial denouement. It is acknowledged that the Indian judiciary has "adopted a fairly interventionist role in moulding telecom regulatory policy in India since the early 1990s" (Thiruvengadam and Joshi 2012: 2). The current and future implications of judicial intervention in telecom regulation on the relationships among the executive agencies concerned and other actors and the 'orderly growth' of the sector deserve closer attention.

1.2 Research framework

A more complex political, economic, and legal institutional matrix than is apparent at first sight undergirds regulation. In the specific context of the telecom sector, technological issues introduce an additional level of complexity that uniquely influences the way regulation is conducted. Different sets of actors – the state in its legislative, judicial, and executive roles; the IRA; the regulated entities that are market actors in the main; international bodies; policy networks; and the vast mass of consumers who overlap the voting population – exert influence at the level of the institutional environment (where the rules of the game are set) and the institutional governance (where the game is played). It is necessary to investigate both sets of activities to arrive at a comprehensive understanding of the structure and functioning of regulatory governance (the state of play).

The present work seeks to examine the Indian telecom regulatory case at three levels: (1) at the macro-institutional level, where it is proposed to study the norms and rules reconstituted over time and analyse whether such a reordering of the rules of the regulatory game has helped or harmed the achievement of stated regulatory objectives; (2) at the level of the institutional environment by taking up the impact of important court judgments; and (3) at the micro-institutional level by examining the mechanisms of governance of TRAI and the manner in which its functioning has affected the alignment of incentives in the regulatory space. The new institutional economics (NIE) framework represented in Table 1.1 may be seen as indicating the broad setting for the study.

We recognize with Black (1997) that, apart from the micro-economic foundations of game theory and transaction costs, the NIE approach is based on the behavioural theory of bounded, not classical, rationality and the idea that an individual's actions are necessarily structured by institutions. It is also true that NIE views preference formation as exogenous, as being outside the institutional context, in contrast to

Table 1.1 NIE framework

Level		*Frequency (years)*	*Purpose*
L1	**Embeddedness:** informal institutions, customs, traditions, norms, religion	10^2 to 10^3	Often non-calculative: spontaneous
L2	**Institutional environment:** formal rules of the game – esp. property (polity, judiciary, bureaucracy)	10 to 10^2	Get the institutional environment right; first order economizing
L3	**Governance:** play of the game – esp. contract (aligning governance structures with transactions)	1 to 10	Get the governance structures right; second order economizing
L4	**Resource allocation and employment** (prices and quantities; incentive alignment)	Continuous	Get the marginal conditions right; third order economizing

L1: social theory; L2: economics of property rights/positive political theory; L3: transaction cost economics; L4: neoclassical economics/agency theory

Source: Williamson 2000.

non-economic new institutionalisms that consider institutions as shaping "not just the actions of individuals and organizations, but their preferences as well" (ibid: 64). We also agree with Black that it would be analytically more sophisticated to move beyond the dichotomy of rational versus contextual actor by positing an 'institutionally rational' actor who acts strategically in pursuit of endogenously formed interests such that, while providing choice on ways forward, institutions provide socially validated standards – that is, the social context in which individual action can be explained. Thus, our conception of new institutionalism seeks to "explain characteristics of social outcomes on the basis not only of agent preferences and optimizing behavior, but also on the basis of institutional features" (Shepsle 1989: 135), and to examine ex ante selection of rules (institutional selection) as well as their ex post robustness (institutional maintenance), *and*, in addition, sees institutions as providing "exterior normative constraints on actions [and] also shap[ing] interior cognition" (Black 1997: 68). In so doing, it recognizes that students of equilibrium institutions, such as economic historians, must "delicately straddle the boundary between micro and macro" (Shepsle 1989: 145).

1.3 Regulatory theory

The polysemic nature of regulation leads to the difficulty in precisely defining the term. Conceptions of regulation range from the broader view that sees it as encompassing all forms of social control to its narrow construal as a specific set of commands or a binding set of rules administered by a specially constituted body, usually by the state through a public agency. The concept of a regulatory state that relies on law-backed specialized agencies with technical expertise that are insulated from quotidian political pressures to effectuate economic decisions has gained ground, as has the argument that regulation arises from the inadequacy of standard contract law to administer special contracts that may require hybrid institutional forms combining aspects of executive, legislative, and judicial functions.

A. Market failure and regulation

Any attempt to arrive at a definition – even a working definition – of regulation must straightaway confront the question of whether regulation is at all 'possible' in the first place. In the economic realm, state intervention is primarily justified on the grounds of market failure. The ability of the state to regulate has been questioned on the basis of the state's propensity to fail on a variety of grounds based on pluralist, elitist, public choice, institutionalist, and poststructuralist grounds. There is now some consensus across the ideological spectrum that the state continues to have a legitimate role in regulation, albeit in a 'decentred' way. The decentred perspective consists of a mélange of approaches, provides an alternative diagnosis for 'failures' of state-centred action, and requires us to "reconsider not only how the state might act in order to pursue its goals, but how we should understand 'regulation' itself" (Black 2002b: 4).

Welfare economics postulates that, under certain assumptions, "the equilibrium conditions which characterize a system of competitive markets will exactly correspond to the requirements of Paretian efficiency" (Bator 1958: 351). Such a general equilibrium will hold only if all markets are perfectly competitive, which of course is impossible in practice, leading to what has been termed 'market failure'. This failure could be of many kinds: (1) monopoly, natural monopoly, and abuse of dominance or market power; (2) externalities or spillover effects; (3) public goods; and (4) information inadequacies and asymmetries. The market failure rationale for regulation could also extend beyond these four broad instances to (1) curbing windfall profits; (2) ensuring continuity and availability of service; (3) restricting anti-competitive behaviour and predatory pricing; (4) correcting unequal bargaining power; (5) managing scarcity and rationing; and (6) avoiding excessively high transaction and coordination costs (Baldwin et al. 2012).

Competition is crucial if general equilibrium is to hold. The first kind of market failure occurs when competition is affected due to monopoly, and government regulation to prevent monopolistic behaviour is considered "least controversially" as having a "facilitative or constitutive function" in order to ensure a well-functioning market (Sunstein 1990: 48). Natural monopoly is a special case of monopoly and recognizes that "the technology of certain industries or the character of the service is such that the customer can be served at least cost or greatest net benefit only by a single firm (in the extreme case) or by a limited number of 'chosen instruments'" (Kahn 1971: 2). In the extreme case, the marginal and average costs of a single firm's production continue to decline until it becomes apparent that a single firm can supply the total output of an industry more cheaply than multiple firms. Utilities such as local telephone companies are often cited as examples of natural monopoly given that they find it progressively cheaper to supply extra units of service due to economies of scale. While increasing competition through recourse to antitrust or competition law may be sufficient to correct for a monopoly situation, this may not be enough to correct for lower output and higher prices in the case of a natural monopoly. The natural monopoly argument has also been used in the theory to justify regulation on the grounds of prohibiting duplication of facilities and curbing windfall profits.

Goldberg (1976) provides the perspective of the 'administered contract' to examine regulatory institutions, whereby the justification for regulation rests on grounds far broader than natural monopoly arguments of declining long-run average costs. He argues that viewing the economic contract in discrete transactional terms crucially ignores two elements of contract: (1) an ongoing relationship between contracting parties and (2) individuals' reliance on agents for gathering information, negotiating contracts, and so on (in short, agency). It is his contention that "at the stage at which contracts are formed, the agent (regulator) must establish in the contract some measure of protection both for the producer's right to serve and the consumer's right to be served", thereby making a trade-off between producer and consumer interests (Mitnick 1980: 305–6). Within the franchise bidding context, it is argued that in cases where the successful performance of contract requires capital investment of long gestation, the future availability of market becomes crucial, leading to a requirement for restricting consumers' ability to exit. Consumers too might desire an assurance of continued service before locking into expenditures on complementary equipment. Since the "protection of the producer's right to serve will make the contract more attractive to producers but, *ceteris paribus*, will make it less attractive to consumers", and vice versa, the agent must determine the appropriate protection between both parties (Goldberg 1976: 433). Hence comes the rationale for regulation where competition does not arise from considerations of natural monopoly, "but from the need to manage long-term relationships, i.e., contracts, among the parties in question" (Mitnick 1980: 307), in the collective interests of societal groups.

Williamson too takes up the ex ante franchise bidding as regulation by drawing attention to the factors that affect the assessment of alternative modes of organizing natural monopoly services, including costs of ascertaining and aggregating consumer preferences directly, the efficacy of scalar bidding, the degree to which technology is developed, demand uncertainty, the degree to which incumbent suppliers acquire idiosyncratic skills, the extent to which specialized equipment is involved, and the "susceptibility of the political process to opportunistic representations and the differential proclivity, among modes, to make them" (1985: 328). He views "regulation as a long-term behavioural relation that structures outcomes" (ibid: 304–5) and states:

> At the risk of oversimplification, regulation may be described as a highly incomplete form of long-term contracting in which (1) the regulatee is assured fair rate of return, in exchange for which (2) adaptations to changing circumstances are successively introduced without the costly haggling that attends such changes when parties to the contract enjoy greater autonomy.
> (ibid: 347)

B. Other forms of market failure

Regulation arising from market failure is also justified on the grounds of curbing windfall profits or economic rents that arise when a firm "finds a source of supply significantly cheaper that that available in the marketplace" (Baldwin et al. 2012: 17). Rent is different from monopoly profit in that the cheap source cannot supply the entire market, and inefficiency, or allocative waste, cannot be assumed in all cases where it is present (Breyer 1998). Regulators may have to find a balance between incentivizing new efficiencies, products, or markets by allowing firms to retain excess profits earned through these mechanisms and appropriating for the larger public such profits that arise merely from good fortune. Especially in the case of utilities, it has been asserted that a change in prices should be effected by regulatory intervention to reflect obsolescence of prices due to improvements in technology. However, even this justification for regulation has been contested on the grounds that uncertainty is not a phenomenon unique to natural monopolies, that the correct way to view the problem is one of selecting the best type of contract, and that long-term contracts "for the supply of commodities are concluded satisfactorily in the market place without the aid of regulation" (Demsetz 1968: 64). In this view, it is not clear whether regulation handles the uncertainty-rent problem better than the market. Regulation to control rent is actually undertaken for other reasons, such as income transfer or fairer income distribution, and to avoid "the suddenness of price increase rather than the existence of rent" (Breyer 1998: 68).

'Externalities' are neighbourhood or spillover effects, whereby market failure occurs "if a producer's activity imposes costs on third parties that are not reflected (or 'internalised') in the prices which he charges for his products [and] a misallocation of resources results; purchasers of the product do not pay for its true social cost and hence more units of the product are supplied than is socially appropriate" (Anthony Ogus, quoted in Morgan and Yeung 2007: 21). Externalities arise because the costs or benefits of a voluntary market exchange spill over to parties "other than those explicitly engaged in the exchange" (Cooter and Ulen 2012: 39) and could accordingly either be 'negative' (such as pollution from industries) or 'positive' (such as bees pollinating neighbouring orchards). Government subsidies are usually justified as supportive of positive externalities, such as in free public education, but the more common externalities encountered in the regulatory world are of the negative kind. Rearrangement of private property rights and application of liability rules to allocate the burden of paying for spillover costs have been suggested as alternative approaches to government regulation, but given the propensity of economic actors to 'hold out'[7] or 'free ride'[8] during the process of bargaining, and the problems inherent in identifying the 'least cost avoider' of costs, government intervention is considered the desirable form of administering externalities.

A third kind of market failure relevant for justifying government regulation is the existence of what are called public goods. Public goods, first formalized in the work of Paul Samuelson, are characterized by non-rivalrous consumption and non-excludability, leading to both free rider problems and problems in determining the willingness to pay: because the consumption of public goods by one individual does not result in a reduction of its availability to others, there is an incentive to free ride, and since it is impossible or too expensive to exclude individuals from consuming a public good even if they do not pay for it, "those with high valuations will tend to understate their preferences in the hope of being charged a lower price" (Veljanovski 2010: 21). The market may be unable to encourage the production of adequate levels of goods or services providing social benefits such as defence, education, health care, or security, and it may be required to regulate their provision through the use of either subsidies or even direct government provision of public services. However, in practice, public goods are not always provided by the state, and even then, not exclusively.

Inadequate information, or 'information asymmetries', is the next class of market failure that may be used to justify regulatory intervention. If markets are working, it is expected that as much information about prices, quantity, and quality of goods would be available as consumers are willing to pay for "in order to lower the cost or to improve the quality of their choices" (Breyer 1998: 72). The market, however, may fail to produce adequate information; asymmetric distribution of information may occur due to many reasons: (1) the high costs of producing information that is very cheap to make available once produced; (2) deliberately misleading information that conveys false information or by omits significant information; (3) information that is hard to evaluate and is therefore of not of adequate utility to the consumer; and (4) collusion in the marketplace leading the generation of inadequate information (Baldwin et al. 2012). Information asymmetries give rise to problems of 'adverse selection',[9] which occurs when goods of low quality drive out goods of higher quality, or 'moral hazard', which "describes a situation in which someone other than a buyer pays for the buyer's purchase" (Breyer 1998: 79, footnote omitted), where parties misuse their information advantage (den Hertog 1999). Although markets have developed signals to counter adverse selection problems (such as advertising and other techniques of product management like brand-name building) and self-regulatory instruments to overcome moral hazard (such as through adherence to professional codes or guild-wide norms), government intervention is deemed justified at least to ensure that adequate information is available in the public domain.

C. Defining regulation

This introductory survey of some market failure arguments and cases is meant to serve as a point of entry into the two broad varieties of theory concerning regulation: (1) public interest theories and (2) private interest theories (Mitnick 1980; den Hertog 1999). A third variety of 'institutionalist' theories, also identified in the literature,[10] "explicitly base their entire approach upon the potential fluidity of boundaries", not only between public and private interest theories but also between normative (or prescriptive) and positive (or explanatory) motivations for regulation (Morgan and Yeung 2007: 17). All these groups of theories build upon the notion that central ordering of economic activity by the state is the subject matter of regulation and essentially debate the desirability and scale of such regulatory activity. It would therefore be apposite at this point to consider definitions of regulation that illuminate further discussion of its theoretical underpinnings. It has been said that "regulation is the employment of legal instruments for the implementation of socio-economic policy objectives" (den Hertog 1999: 223) and that regulation can be understood "essentially as state intervention into the economy by making and applying legal rules" (Morgan and Yeung 2007: 16). A trifle more generally, but still amenable to practical use, are definitions that refer to regulation as "the promulgation of an authoritative set of rules, accompanied by some mechanism, typically a public agency, for monitoring and promoting compliance with these rules" (Baldwin et al. 1998: 3) and as "the public administrative policing of a private activity with respect to a rule prescribed in the public interest" (Mitnick 1980: 7). While legal instruments or rules are common to all three, the first definition is agnostic to the presence of the state as implementer of regulation, the second is explicit about the state's role, and the last two assign some sort of primacy to the state in typical regulatory settings by recognizing the mechanism of the public agency for policing and achieving compliance. In setting boundary conditions to arrive at a definition, it is necessary to explicitly recognize the economic nature of regulation that concerns the present study of telecom regulation. It is also required to move beyond static views of regulation so the structure and dynamics of the intentional restriction of choice it engenders is taken on board; in particular, "regulation may be viewed as a form of *agency* relation, i.e., a relation in which an *agent* is acting for a principal" (ibid: 9, emphasis in original). The regulator as principal has to restrict the choice of activities of the regulated entity as agent by using incentives or directives that may be attended by agency problems. For the purpose of the present study, which intends to focus on telecom services regulation in the context of the Indian state, the following working definition of regulation is proposed to be adopted for further theorizing about the concept:

> Regulation is the process by which the state authoritatively intervenes in the economy and manages long-term relationships among parties by promulgating or modifying legal rules, putting in place incentives, and creating or identifying appropriate public agencies to coordinate implementation of the rules and monitoring compliance.

This definition is only the first step in theorizing regulation; indeed, it could even be suggested that any definition should *follow* from a balanced evaluation of the different theories of regulation, especially since the working definition adopted here may be seen to manifest a bias for the normative, public interest conceptualization of the term.[11] Public interest theories of regulation are normative in the sense that they tend to view regulation as justified based on efficiency and market failure considerations and prescribe government regulation as an antidote to the failure of the market. As can be expected, there are a wide variety of views regarding what constitutes public interest across public interest theorists and theoretical viewpoints.

D. The public interest

A good starting point in identifying public interest is the economic, where public interest can be construed as "the best possible allocation of scarce resources for individual and collective goods" (den Hertog 1999: 225). In this view, public economic interest is substantially achieved if (1) transaction costs can be reduced by enacting and implementing stable rules regarding property and contract and by rectifying information asymmetries and external effects as they occur, (2) market imbalances are corrected to avoid excessive competition that may be allocatively inefficient, and (3) free rider problems in public goods' supply are tackled, all three by government intervention. The economic standpoint attributes substantial importance to the abatement of market failures, and government regulation is seen as the alternative, if second-best, solution to the practical problems encountered in making markets work efficiently. Researchers have sought to correct the predominantly economic slant of this view by importing values other than those concerned with achieving allocative efficiency, and these have been described as 'substantive political' and 'procedural political' public interest theories of regulation (Morgan and Yeung 2007). Political versions of public interest theories are seen to adopt a more multi-faceted conception of public interest arrived at "by deliberation, mutual interchange, dialogue and collective processes" (ibid: 27).

In substantive political approaches, market failure justifications are supplemented by other defects in market ordering, with due recognition for the view that the "choices people make as political participants are different from those they make as consumers" (Sunstein 1990: 57). Relying on civic republican notions of virtue (Morgan and Yeung 2007), the range of non-economic substantive public interest goals of regulation is expanded to include (1) public-interested redistribution of resources from one group to another, (2) attempts to foster and promote diverse experiences, (3) elimination or reduction of social subordination of various social groups, (4) shaping of endogenous preferences, and (5) solving the problem of irreversibility of certain courses of conduct.

Procedural political approaches stress the deliberative aspects of dialogue between participants to decide among alternative regulatory pursuits such that constraints are placed on regulatory procedures to reduce power inequalities. Although different from civic republican traditions of theorizing public interest, proceduralist approaches could be classified as 'thin' or 'thick', where the thin conception involves the participation of citizens "in which preferences remained exogenous, unchanged, and which was discourse-less", and thick proceduralization "is based on deliberative democracy and concepts of rational discourse derived from Habermas" (Prosser 2006: 377, footnote omitted). In Prosser's view, providing a space for ethical and moral discourse concerned with questions like "what kind of society do we want to be?" and "what is good for us all?" allows regulation to address broader substantive concerns sidelined by the welfare economics approach in regulatory studies. This approach critiques the view that regulation is a response to market failure since that means accepting it is second best to the market mechanism and instead views regulation (and laws) as serving "to constitute market relations" (ibid: 382). To avoid the charge of excessive procedural orientation, this approach goes on to incorporate notions of citizenship and inclusivity and social solidarity in the Durkheimian tradition so as to limit the socially fragmenting role of the market.

This discussion of the kinds of public interest theories of regulation also contains the seeds of the different criticisms levelled against them. One potent criticism is the limited explanatory power of the theory of market failure often used to justify welfarist notions of regulation in the public interest, as is the overly optimistic assessment inherent in public interest conceptualizations that government intervention can cure market failures. Another is the view that they "do not distinguish origin explanations from operation explanations" (Mitnick 1980: 105). It has also been pointed out that it is wrong to assume regulatory intervention is costless. It has been argued persuasively that before proceeding to declaim the benefits of government intervention over private enterprise in correcting divergences between social and private costs, it is desirable to "compare the total product yielded by alternative social arrangements" (Coase 1960: 43) without assuming government regulation is ideal and costless.

It has nevertheless been noted that people "genuinely believe the rhetoric that regulation is supposed to protect society from market failures" (Noll 1983: 403) and that government intervention, while not perfect, is the most efficient way of tackling market failure since government has the ability to obtain information less expensively. Notwithstanding this development, there has been a strong tendency in the literature to debunk public interest versions of regulatory theory. The argument that regulation is supplied by the state in response to a demand arising from failures or inefficiencies of the market has been seen as deficient from many different viewpoints. As may be expected, these are commonly listed under the rubric of private interest theories of regulation.

E. Private interests in regulation

Private interest theories are based on the assumption that regulation emerges from the action of self-interested, interest-maximizing individuals or groups and that any connection between regulatory intervention and the public interest is a contingent one (Morgan and Yeung 2007). Private interests have free play in shaping regulatory outcomes, and unequal distribution of resources could well be reflected in differential advantage accruing to more powerful groups or individuals, irrespective of public interest. Two broad varieties of private interest theories can be discerned, depending on whether the primary disciplinary lens is political or economic. Political economy private interest approaches are basically derived from the pluralist and neopluralist versions of state theory discussed earlier, with an emphasis on interplay of group interests with minimal central ordering. Economic theories of private interest have built upon critiques of what has been termed 'capture theory'[12] in the literature, mainly developed by political scientists (den Hertog 1999), and are marked by their manifest opposition to public interest justifications and the ability of regulation to correct market failures. While acknowledging that political scientists have developed some evidence of the importance of interest groups in the formation of public policy and in the working of legislative and administrative processes – as reflected in the formulation of capture theory, which postulates that regulatory agencies come to be dominated by the regulated industries with the passage of time – economic theories of private interest insist that such work "does not tell us why some interests are effectively represented in the political process and others not, or under what conditions interest groups succeed or fail in obtaining favorable legislation" (Posner 1974: 341, footnote omitted). They seek to fill the gap with private interest theorizing.

Economic theories of private interest seek to integrate "the analysis of political behavior with the larger body of economic analysis" (Peltzman 1998: 93), drawing upon early twentieth-century pioneering work in by Arthur F Bentley, who used an 'economic approach' that focused on political pressure groups, and followed later by other pluralist political scientists such as Truman (Becker 1983). Its origins reflect a felt unease with the lack of empirical proof for the many assumptions of public interest theory, a suspicion that "innumerable regulatory actions are conclusive proof, not of effective regulation, but of the desire to regulate" (Stigler and Friedland 1962: 1), and a sense that, when empirical cases are investigated, as for example in the regulated industry of electricity production and distribution, regulation proves ineffectual in practice. The 'Chicago School' theory of regulation, inaugurated famously by Stigler's paper in 1971, notes the utility of the coercive power of the state for industry and has as its central thesis the notion that "as a rule, regulation is acquired by the industry and is designed and operated primarily for its benefit" (Stigler 1971: 3). Along with its insistence "with political scientists that economic regulation serves the private interests of politically effective groups" (Posner 1974: 343), it offers a general framework that emphasizes the self-interested political behaviour of politicians (who, for simplicity, are assumed to direct the behaviour of regulators), as well as constituents who exchange objects of utility, and the importance of information and organization costs that constrain a group's ability to deliver votes and money. Stigler's main conclusions from an analysis of costs are that "producer interest will win the bidding for the services of a regulatory agency" and that, more generally, "in any similar political contest between groups of disparate size, the compact organized interest (say, farmers in a developed economy) will usually win at the expense of the diffuse group (taxpayers)" (Peltzman 1998: 98).

The theory benefited from internal critiques, most notably from Posner and Peltzman, and also from its extension by Becker (Peltzman 1998). Posner critiqued private interest theory on, inter alia, grounds of excessive generality and insufficient or unsystematic empirical evidence but concluded that "the success of economic theory in illuminating other areas of nonmarket behavior leads one to be somewhat optimistic that the economic theory will eventually jell" (1974: 356). His criticism also served to question the generalization that regulation served producer interest by drawing attention to the phenomenon of 'internal subsidization' or 'cross subsidization' whereby there is "enforced provision of service to selected consumer groups at specially low, often below-cost, prices . . . 'subsidized' out of potential producer rents generated elsewhere in the regulated industry" (Peltzman 1998: 99). This notion was extended by Peltzman to posit that "no single economic interest captures a regulatory body" (ibid) by viewing regulatory activity as a constrained maximization problem wherein the regulator is constrained by the demand and cost functions of the regulated industry from maintaining a politically optimum distribution. His approach, which has been criticized among others as not being a model of regulation but of legislation-making, "is essentially a model of wealth redistribution by elected politicians seeking to maximize votes" (Mitnick 1980: 121). Becker contributed to further fine-tuning of the theory by introducing into the analysis the concept of 'pressure groups' whereby individuals belonging to particular groups (such as by occupation, income, etc.) are assumed to use political influence to enhance the well-being of their members, and "competition among these pressure groups for political influence determines the equilibrium structure of taxes, subsidies, and other political favors" (Becker 1983: 372). By further assuming that taxes, subsidies, regulations, and other political instruments are used to raise the welfare of more influential pressure groups, but appreciating that there are limits to transfers of income from more efficient pressure groups to less efficient ones due to welfare losses or 'deadweight costs', he concluded that policies that raise efficiency are more likely to win the competition for influence because they produce gains rather than deadweight costs, giving the beneficiary groups an intrinsic advantage over groups that are harmed in the process. He thus made allowance for the market failure proposition as well by showing that it too "is produced by competition among pressure groups for political favors" (ibid: 396).

The Chicago theory has been criticized as incomplete on the grounds that it does not pay attention to (1) the motivation and behaviour of various political actors, (2) the interactions between the various actors in the process of regulation, and (3) the mechanism through which regulators conform to the wishes of the organized interests, as well as for its assumption that legislators always fully honour the wishes of interest groups, without accounting for attendant 'principal-agent' problems (den Hertog 1999). Using the concept of 'subgovernment', Weingast notes from the social choice theory perspective the *simultaneous* influence of many groups in a way not anticipated by Peltzman, and asserts that where many different interests compete over a single policy domain, the difficulty of changing the status quo would increase, and "policy would represent a balance of these interests based on the relevant political returns generated by each faction" (1981: 175). Chicago theory has also been criticized from the viewpoint of public choice theory, which, as noted in the earlier section on the state and its institutions, is focused on pursuit of private gain and considers group interests secondary to or consistent with such a pursuit. State power is monopolized by the powerful few who use it for their own benefit. The "Virginia School of Public Choice" (Buchanan, Tullock, Niskanen) criticism of private interest theory arises from a view that rent seeking (whereby interest and pressure groups pursue monopoly rights granted by the state) leads to inefficient use of scarce resources on the part of the interest groups as well as by potentially disadvantaged consumers who would seek to block the creation of a monopoly if possible (ibid). In conclusion, while private interest theories have explanatory rather than normative value and are useful for the scepticism they bring to overly optimistic assessments of state capacity to overcome market failure, they do little justice to law and regulation – or indeed to regulatory reality as it exists around the world today – by limiting their potential utility to merely constituting the market and no further (Morgan and Yeung 2007).

F. Responsive regulation and regulatory space

We turn next to institutionalist views. What have been termed 'institutionalist' theories of regulation are in fact a mixed bag of very different theories but are, however, distinguished by a common adherence to "rule based spheres, or the relationship between different rule-based spheres" such as formal organizations, embedded norms and routines, or 'systems' as understood by system theory (ibid: 53). Notwithstanding quite substantive differences between them, these theoretical approaches are united by a concern with institutional dynamics and the blurring of differences between public and private actors and interests. They reflect a movement in regulation theory "away from a pure interest-group driven analysis, towards a growing emphasis on institutional design and a coupling of this with a more detailed differentiation of motivations and behaviours as these are encountered in the body of politicians, regulators, and the regulated that makes up the regulation community" (Baldwin et al. 2010: 11).

The first among this ensemble of institutionalist theories could be termed 'republican social-democratic', incorporating the ideals of deliberativeness, political equality, universality, and citizenship that Ayers and Braithwaite (1995) have termed 'responsive regulation'. It seeks to reconcile the divergent positions between state regulation and private ordering that increasingly came to be seen in 'either/or' terms in the Reagan-Thatcher years, when deregulatory impulses were accelerated in the United States and Britain and both public interest theories and private interest explanations of regulation tended to dissolve in the face of developments in the "economic regulation of sectors such as transport (airlines and freight), telecommunications, energy and the financial sector" (den Hertog 1999: 245). Responsive regulation, without being programmatic or prescriptive, seeks to embed regulatory strategy in context, regulatory culture, and history and suggests that regulation be responsive to industry structure and to industry conduct and that government "be attuned to the differing motivations of regulated actors" by progressively and effectively delegating government regulation of the marketplace to public interest groups, to unregulated competitors of the regulated firms, and even to the regulated firms themselves (Ayers and Braithwaite 1995: 4). In this view, "responsiveness means having respect for the integrity of practices and the autonomy of groups" (Braithwaite 2006: 885), and, in contrast to the control imperative, regulation should entail "respect for and deference to the system of private ordering" (Selznick 1985: 363). To achieve the objective of responsiveness while at the same time avoiding the risk of capture and corruption potentially inherent in the cooperative process between regulatory agencies and industry, the concept of 'tripartism' – in which relevant public interest groups become the third player in the regulatory game to counterbalance the regulated actors as well as to make the regulation (or the guardianship activity) contestable – is introduced. Public interest groups are at the cutting edge of a contested, democratic theory of public interest, and the institutional design of tripartism "can create a system of checks and balances that harness private interests to work in favour of the public interest" (Morgan and Yeung 2007: 59). Responsive regulation could thus be viewed as enforced self-regulation, which stands at the middle tier of the 'enforcement pyramid' of regulatory strategies involving self-regulation at the bottom tier and state regulation with discretionary punishment at the top, with each higher tier being successively adopted if the one below fails (Baldwin et al. 1998). The theory of responsive regulation has been influential in explaining the evolution of regulatory regimes in developed countries, but also has some resonance for developing countries such as India. While recognizing that developing countries may not have the same responsive regulatory capacity as wealthier ones, it has been asserted that these capacity deficits can be overcome – and a regulatory society created by effectively bypassing the regulatory state – by using "strategies of networking around state incapacity and mobilizing private markets for enforcing virtue" (Braithwaite 2006: 896) so that a sort of 'efficient capture' is enabled as deliberative processes become compatible with cost-benefit payoffs.

A second type of institutional theory of some interest in the Indian context is that of 'regulatory space', which "emphasises a place where regulation occurs, almost a kind of physical arena which influences the practices that happen within it" (Morgan and Yeung 2007: 59). This theory defines the analytical construct of regulatory space by the range of regulatory issues subject to public decision and seeks to analyse the ways in which "national political and legal settings, historical timing, organizational structure, the character of markets, and the nature of issue areas all influence the shape of regulatory space and the allocation of power within that space" (Hancher and Moran 1998:

148). It is a "furiously contested" space (ibid: 153) that "contains not only state actors and formal public authority, but also non-state actors and sources of authority over which the state may not have a monopoly such as information, wealth and organisational capacity" (Morgan and Yeung 2007: 60). This approach has much to offer "simply because it does not privilege any particular relationship in explaining or predicting how regulation will operate" (Prosser 1999: 213). It is only to be expected that the concept of 'space' implies inclusions and exclusions. An important insight of this theory that has particular applicability in the Indian context is that, given the propensity of large organizations that dominate regulatory relationships to be organized in complex administrative hierarchies, "routine application of established practices, rather than individual choice, is the dominant influence in deciding who is taken into, or kept out of, regulatory space", thereby making exclusions systematic (Hancher and Moran 1998: 154). Another crucial insight relates to the importance of national traditions in populating the respective regulatory space with different actors, especially of culturally formed assumptions about the purpose and role of law; regulatory space theory asserts that "the character of a legal culture mediates the regulatory process, fixing the scope of regulatory space and influencing who gains entry and on what terms" (ibid: 158). Other important ways in which the theory has contributed to regulatory studies is by drawing attention to the influence of historical timing (how institutional forces are poised at critical moments in regulatory history), of organizational structure (especially organizational subcultures crucially influenced by forms of organization of state and rule-making bodies), and of the emergence and development of networks (how interdependence between organizations is institutionalized). The regulatory space perspective – which has been further extended to cover what are termed 'regulatory regimes' – goes beyond capture theory "to examine the process by which powerful organizations, both state and non-state, gain, maintain, and sometimes lose their dominant positions within regulatory space" (ibid: 171). In doing so, it reduces the "relevance of law and formal public authority" (Morgan and Yeung 2007: 60) within that space, with important implications for the study of the way in which telecom regulation in India has evolved in practice.

G. Regulatory techniques

Having briefly traced the outline of theories of regulation, we now turn to a sketch of regulatory instruments, styles, and techniques as they have evolved over time. It is necessary for a proper appreciation of regulatory styles to go beyond national variations since regulatory approaches and attitudes can vary across individual sectors *within* particular states. Further, the polycentricity of regulation observed in practice mandates proper appreciation of the differential nuance in the techniques applied by IRAs, government departments, and other regulatory bodies such as competition commissions and supranational organizations even with regard to particular regulated sectors. The regulator does not stand in isolation in applying regulatory technique to the objective at hand, and since institutional credibility is usually enhanced if a coordinated approach is adopted by different regulatory actors, additional complexity is introduced into the analysis of regulatory style and technique by the different directions in which it is pulled by (1) regulatory techniques perceived as applicable across jurisdictions and sectors and (2) regulatory styles in sector- and nation-specific institutions embedded in particular societal contexts. The cultural viewpoint takes organizational form and culture as the starting point of such analysis. One way to look at the issue is to construct a matrix of strong and weak 'group' and 'grid' to differentiate different kinds of society where "group refers to the extent to which individuals are integrated into collectives and grid refers to the extent to which transactions between individuals are governed by general rule rather than every transaction needing its own negotiations and *ad hoc* bargaining" (ibid). Such a typology yields different modes of regulation that may preponderate in different kinds of society, as reflected in Figure 1.1.

Another way of classifying regulatory styles and instruments is based on the underlying 'modality' through which behaviour is sought to be controlled, and five classes are identified as part of a heuristic typology: command, competition, consensus, communication, and code (or architecture) (Morgan and Yeung 2007). The first of these, involving rule-based coercion, is also referred to as 'command and control' (CAC) regulation, or 'classical' regulation, and is pithily described as "regulation by the state through the use of legal rules backed by (often criminal) sanctions" (Black 2002a: 2) and "involves the regulator making and enforcing the rules" (Baldwin et al. 1998: 24). In a sense, CAC is set up as the straw man to knock all that is considered bad about regulation and centrally controlled economies, including "poorly targeted rules, rigidity, ossification, under- or over-enforcement, and unintended consequences", and its failings include "instrument failure", "information and knowledge failure", "implementation failure", and "motivation failure and capture theory" (Black 2002a: 2–3). CAC is also 'classical' since this is how regulation

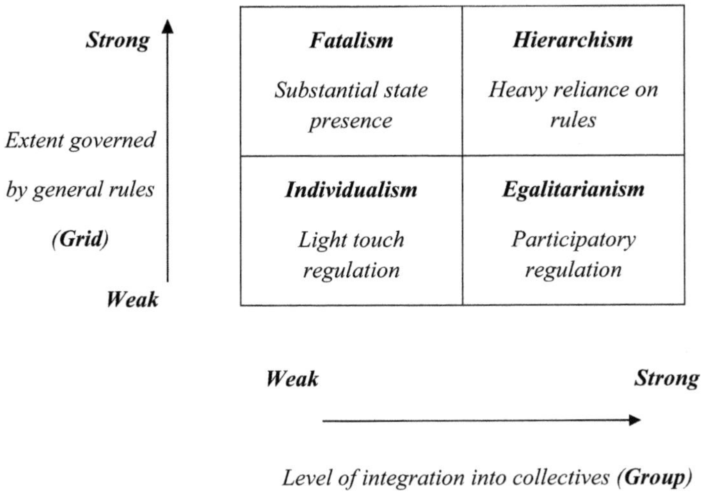

Figure 1.1 Modes of regulation in different kinds of society
Source: Adapted from Baldwin et al. 1998.

was traditionally conceived before the challenges of the theories of private interest, decentring, and various forms of co-regulation in practice undercut the basis as well as the practical effectiveness of command and control. The primary criticisms of CAC have arisen from (1) neoclassical liberalism with its emphasis on markets and prices, (2) rational decision-making models in their application in policy analysis with their sense of awe for the workings of market-like mechanisms for making collective decisions, (3) Skinnerian critiques of coercion and punishment as inefficient in influencing behaviour with a preference for 'operant conditioning', and (4) increasing scepticism about the reliance on coercive approaches even as the scale and scope of government activity burgeoned over time (Neiman 1980). One way in which the negative effects of CAC could be mitigated is by using less prescriptive rules, such as competition rules intended only for dominant players, as have been adopted in the liberalization of telecom markets in the United Kingdom and elsewhere in the world (Baldwin et al. 1998). While it is true that command-based mechanisms of regulation have generally fallen out of favour, there is also some scepticism of the wholesale debunking of CAC especially from a political science viewpoint that warns against underestimating the potential benefits and functions of coercive approaches to policy implementation and, in contrast to those "who view the regime as a mere adjunct to the marketplace, [views] the market and the system of exchange as a fact and product of political will" (Neiman 1980: 32). Command approaches may still retain applicability in some regulatory circumstances, and as with other generalizations, an outright rejection of CAC as outmoded needs to be attuned to differential capacities among different kinds of regulated risk-creators and "made sensitive to differences in behavioural response as regards different actors in different regulatory environments" (Ogus 2009: 346).

Competition-based regulation, in contrast to CAC, seeks to "harness the competitive forces arising from rivalry between competing units as a means for regulating social behaviour" (Morgan and Yeung 2007: 85) so that the law comes to play a facilitative role rather than a coercive one. Competition could be induced through a variety of economic instruments, such as charges and taxes, subsidies, and – to give an example from environmental regulation – tradable emission rights. Taxes could be used to increase the cost of certain activities and could be levied to incentivize carrying out activities at economically efficient levels by ensuring that "the costs of negative externalities are really paid by those creating the externalities", while subsidies could be employed "to alter the incentives faced by individuals or organizations" (Baldwin et al. 1998: 27). It must be noted, however, that the coercive power of regulation (or law) continues to exert an important influence in such an approach. This coercion could happen in two ways: (1) law's command could operate in an indirect manner by backstopping competitive instruments (Morgan and Yeung 2007) and (2) the putative reward could be *perceived* as punishment in disguise by the regulated entity (Neiman 1980).

Another technique that could be employed is "to shape social behaviour by altering the legal liability associated with particular conduct", which could be based on the competitive force of markets (Morgan and Yeung 2007: 88). The general idea is to construct liability rules in such a way as to ensure that the costs of an activity are placed on the party in the best position to weigh the costs against the benefits. The following steps, or series of arguments, abstracted from Calabresi and Melamed (1972), conveys the logic behind the thinking: (1) Pareto optimality would occur regardless of initial entitlement in the absence of transaction costs, but transaction costs are always present in practice; (2) economic efficiency dictates that entitlements should reflect knowledgeable choices made between obtaining and avoiding social benefits and social costs; (3) since it is uncertain whether a benefit is worth what it costs to society, the cost should be placed on the party (or activity) best located to make the cost-benefit analysis; (4) in some contexts such as accidents or pollution, this would mean putting the costs on the 'least cost avoider', that is, the party that can most cheaply avoid them; (5) where the least cost avoider is not identifiable, the costs must be placed on the party best able to induce others to correct the error or act more safely; and (6) a decision would be required whether market transactions or government intervention are more likely to achieve economic efficiency. However, it has been argued that shifting liability rules is difficult in practice since (1) liability rules embody a complex system of incentives, (2) the court system itself works imperfectly, (3) the shift will affect relative wealth of parties, and (4) the process of changing liability rules may have other undesirable social consequences such as questions regarding retroactive applicability (Stephen Breyer, quoted in Morgan and Yeung 2007).

The next subset of techniques relates to consensus, including forms of self-regulation "through to those involving various forms of co-operative partnerships between state and non-state actors in seeking to regulate social behaviour" (ibid: 92). The first concept of self-regulation encompasses a wide range of institutional arrangements and is often invoked by professional bodies and elite associations of experts to keep out exogenous regulation. The public interest justification for self-regulation is said to be based on three conditions: (1) that the regulated activity is affected by market failure (notably externalities and information asymmetries); (2) that private law is inadequate or too costly to correct the failure; and (3) "that self-regulation is a better (cheaper) method of solving the problem than conventional public regulation" (Ogus 1995: 97). While the first two conditions are common to all public interest theories, the third contention is at the centre of self-regulation. Ogus constructs two models of self-regulation: (1) the first, where no significant externalities or information asymmetries exist, is characterized by Coasean or consensual bargaining, with a largely residual role for public agencies once general regulatory goals are laid down by the legislature; and (2) the second model, which has more general applicability, is called competitive self-regulation and allows consumers to choose between competing self-regulatory regimes, the competition itself being one of three varieties – unrestrained market competition, agency-assisted, or ex ante competition for ex post monopoly – once again with a residual role for IRAs. Hybrid models that stress the influence of both command and consent in achieving regulation have the effect of making the law's threat recede to the background, "although it does not disappear entirely" (Morgan and Yeung 2007: 96).

The technique of communication aims to utilize the force of social norms and consensus and "attempts to persuade and educate members of the regulated community, or those affected by the regulated activity, to act in a manner that will facilitate the achievement of regulatory goals" (ibid) and can be seen as a powerful adjunct to other regulatory techniques. Communications that occur among the regulators, the regulated, and others involved in the regulatory process are of myriad kinds, including "all forms of interpersonal communications, extending beyond standards, policy documents, and guidance notes to include all micro-level conversations that may occur in formal or informal settings, including policy briefings, seminars, and conferences, in the course of the regulatory process between individuals both within and across organizations or particular cohesive communities" (Black 2002b: 171), and could consequently be amenable to 'discourse analysis'. Some of the techniques employed in regulatory communication are (1) mandatory disclosure, a form of CAC where the state compels disclosure with the backing of civil or criminal sanctions; (2) voluntary disclosure; and (3) public communication by the state or regulator, including through public information campaigns, guidance, or publicizing compliance or non-compliance (Morgan and Yeung 2007).

Code, or 'architecture', has come to be seen as one of four regulatory constraints on human behaviour – the other three being the law, social norms, and the market – and is seen to "constrain through the physical burden [it imposes]" (Lessig 2006: 124). Code-based regulatory techniques seek to "eliminate undesirable behaviour by designing out the possibility for its occurrence" (Morgan and Yeung 2007: 102). Regulatory styles based on the concept of code can be attributed to the growth of network environments (especially the internet) and the realization that technological capabilities and system design choices impose rules on participants and that the "set of rules for information flows imposed by technology and communication networks form a 'Lex Informatica' that policymakers must understand, consciously recognize, and encourage" (Joel Reidenberg, quoted in Lessig 1999: 410). This may become crucially important when 'commons' are considered in their relationship with code, for it may happen that the requirement to provide certain services to the public generally in a transparent and non-discriminatory way may be compromised by the newfound ability of service providers to "configure their networks and services so that they will fall outside established definitions and obligations and thereby circumvent common carrier obligations" via what is termed 'policy routing' (Roseman 2003: 99). While there is a continuing debate about the reach of regulation in built environments that are "architecturally designed" and heavily influenced by code (such as cyberspace), there is some merit in the view that the law (and regulation) may be seen as operating in a facilitative and permissive fashion, "although its force may appear almost invisible to those operating within [it]" (Morgan and Yeung 2007: 104).

1.4 Issue areas

Part II of this book aims to chart out the interplay between legislative effort at norm-creation and judicial activity at norm-application during the first stage of judicialization of regulatory governance in the telecom sector. The first few chapters in this part deal with the enactment of the TRAI Act, 1997, and its subsequent amendment in 2000. The chapters analyse the consequences of important court judgments on the regulatory environment and their impact on the amendment of the rules of the game in terms of both how the statute to institute an independent regulator (as the dispute resolver constituting the triad) was first encouraged and how it was modified so the internal hierarchy of regulatory endeavour came to be subtly reorganized by affording enough traction to judicial regulation (and therefore to the doctrine of separation of powers even if not maintained "in its full rigour"[13] in the Indian context) as reflected in separating adjudicatory functions from TRAI and vesting them in a separate specialized tribunal. The subsequent chapters in Part II interrogate one particular subsequent court intervention that checked the dominance of specialization and expertise in the adjudication of sector disputes and privileged a more traditional juridical approach to dispute resolution between licensor and licensee, thereby returning the adjudicatory process to a more general jurisprudential path, albeit with a recognition of the limitations of an over-legal approach to social justice. We argue that such a judicial shaping of policymaking goes beyond merely offering passive support to the deregulatory endeavour of the legislature-executive and is reflective of the trend towards meta-regulation that "encompasses any set of institutions and processes that embed regulatory review mechanisms on a systematic basis into the every-day routines of governmental policymaking, such that a particular form of economic rationality becomes part of the taken-for-granted ways of policymaking" (Morgan 2003: 490). Regulatory governance in the telecom sector in India is in this sense a site where judicialization of the political arena intersected with a non-judicial legality that carved out new institutions steeped in expertise that nevertheless needed to be accountable to the larger society and subject to the rule of law.

Part III focuses on examining how TRAI, as the primary institution of regulatory governance in the telecom services sector, has fared in achieving the NIE objectives of crafting order, mitigating conflict, and realizing mutual gain in its non-adjudicatory role. We treat developments in the sector in three phases: the first between 1997 (when TRAI was established) and 2003 that we have labelled preparatory, the second between 2004 and 2007 when competition came to be established, and the third, ongoing phase from 2008 that was largely marked by hypercompetitive market arrangements.

The preparatory first phase itself can be viewed as being composed of two stages: the first stage being the period between 1997 and 2000 before the TRAI Act was amended and substantive changes effected in the institutional arrangements; and the second stage being the three years after the amendment when the regulatory remit changed. The first stage was marked by bitter turf battles between DoT and TRAI, unsurprising given the long-held reluctance of the executive branch to give up regulatory powers. The Amendment Act of 2000 carried through the categorization initiated at the behest of the judiciary of regulatory functions into adjudicatory and non-adjudicatory, with the first entrusted to a specialized oversight tribunal. The non-adjudicatory functions were divided by statute into those TRAI would give recommendations upon and those it would 'discharge'. The subtle downgrading of the IRA by virtue of the creation of TDSAT was compounded by the absence of any meaningful role for it in licensing matters, spectrum management, and equipment standards. The institutional imbalance between DoT and TRAI thus continued under the amended Act as well, which caused many subsequent regulatory mishaps.

The inadequate regulatory response to important contractual issues was reflected in TRAI's recommendations over time regarding the crucially important spectrum resource. While the ambit of the regulator's recommendatory powers was considerably circumscribed by statute and was institutionally absent during the first rounds of privatization, it is nevertheless true that TRAI was subdued in its regulatory response to the need for robust economic management of spectrum. It did not carefully interrogate the vexed relation between license and spectrum that had evolved in an ad hoc fashion prior to its creation in 1997 and compounded its meekness by limiting itself to exhorting the licensor to address the problem of inadequate availability of spectrum. This meekness and inadequate regulatory response spilled over to crucial decisions taken in the second phase between 2004 and 2007 during which the licensing regime flourished with all its imperfections and TRAI was completely isolated from spectrum policy decisions. Regulatory flip-flop culminating in TRAI's recommendations in August 2007, followed by the rush to hypercompetition with the simultaneous grant of as many as 122 licenses in 2007 to 2008, directly led to the judgment in the 2G Case and the Supreme Court's advisory opinions in the presidential reference that mark the high points in the regulatory journey in the third phase, post-2008. Our examination of their fallout indicates the spectrum management regime has taken a turn towards vestment of greater property rights in the resource among licensees or winning bidders since auctions have become mandatory in the medium term for alienation of spectrum. Whether this is conducive to unlocking the value of this scarce natural resource given the technological developments that make a case for the advocacy of the commons approach to spectrum is an open question. The latest judicial intervention has locked spectrum management into a private-property-rights regime without regard for the arguments on the side of keeping open a modicum of public access to the resource. It is important to acknowledge the exhibited preference for a property rights paradigm over a commons regime inherent in the policy choice made consequent to the intervention. We conclude that choice is likely to complicate the legacy if the case for a commons dispensation is strengthened in the foreseeable future.

The final section (Part IV) interrogates the impact of formal and informal rules at the level of the central regulatory organization taking as given the expectations from a set of regulatory arrangements. The way TRAI is structured in turn has important downstream implications for how the 'regulatory game' is played, and it is our contention that the nature of institutional design has indeed had an impact on the regulatory governance of Indian telecom in the years since TRAI's inception in terms of both directing the nature of the regulator's interventions on the one hand and eliciting responses from regulated entities and the licensor on the other. The chapters in this part explore the organizational issues surrounding TRAI to better appreciate the structural and behavioural aspects of where and how power inheres in the regulatory system – the 'micro-politics' of governance, as it were. The relevance of a micro-level examination of TRAI's organizational structure is important not only to understand its impact on regulatory decision-making but also to impart granularity to the institutional and legal analysis of the regulatory governance framework.

Statutory provisions in the TRAI Act that govern the regulator's structure, eligibility, terms and conditions of its members and staff, and so forth carry the imprint of the inorganic amendment of the Act in

2000 in response to the courts' rejection of vesting adjudicatory functions in the same expert body previously tasked with economic regulation. Without prejudice to the utility of separating the adjudicatory jurisdiction from the regulatory functions (including recommendatory functions), the creation of TDSAT and the consequential changes to the Act gave rise to certain statutory shortfalls, silences, and gaps in the structure of TRAI's organization. We conclude that neither legislative interventions not judicial correctives, let alone executive actions, have resulted in restoring the fine institutional balance that is sine qua non for effective regulatory governance of the Indian telecom sector.

Notes

1 The idea of regulatory governance is borrowed from Levy and Spiller, who define it in terms of "the mechanisms that societies use to constrain regulatory discretion and to resolve conflicts that arise in relation to these constraints" (1994: 205). Unless these mechanisms are successfully incorporated into regulatory design, incentive structures alone are seen to be analytically inadequate in explaining performance.

2 That this was closely preceded by the US Telecommunications Act, 1996, "designed to promote competition and reduce regulatory burdens" (Dossani 2002: 71), is also significant.

3 The NTP 1994 has been criticized as badly written and poorly conceived, with "goals and objectives [that] were overly broad and impossible to achieve" (Raghavan 2007: 76).

4 The new National Telecom Policy 2012 (NTP 2012) was approved on May 31, 2012.

5 The National Digital Communications Policy 2018 is now in place.

6 The year 2000 also saw the establishment of Bharat Sanchar Nigam Limited (BSNL) as the corporatized legatee of the service provider role of DoT.

7 A hold out is a bargaining cost that occurs when "sellers have an incentive to hide their true valuation" (Calbresi and Melamed 1972: 1107) and the market is not successful in establishing it.

8 Free riders are those participants (typically buyers) who systematically underestimate the true value of a benefit and hope to "benefit at no cost to themselves from the payment of others" (Cooter and Ulen 2012: 41).

9 Akerlof explains adverse selection in the following terms: in many markets, buyers use some statistic to judge the quality of prospective purchases, creating incentives for sellers to peddle poor quality goods "since the returns for good quality accrue mainly to the entire group whose statistic is affected rather than to the individual seller", leading to a reduction in the average quality of goods and also in the size of the market (1970: 488).

10 It is important to note that an institutionalist approach is so widely accepted that "it is hard to find anyone who would not claim to be an institutionalist" (Baldwin et al. 2012: 53). However, diverse institutional approaches have one thing in common: they look beyond merely

economistic explanations for regulation and seek answers in institutional structures and arrangements as well.

11 It is to avoid the element of bias that the definition eschews any discussion of the *objective* of regulation, even though the achievement of efficiency could well be incorporated as one.

12 'Capture' refers to the predisposition of regulators "to make decisions and take actions consistent with the preferences of the regulated industry" to become, in effect, the agents of the industry (Mitnick 1980: 207).

13 It has been observed that "our Constitution although does not incorporate the doctrine of separation of powers in its full rigour but it does make horizontal division of powers between the Legislature, Executive and Judiciary" (*Clariant International Ltd. and Anr.* v. *Securities and Exchange Board of India*, AIR 2004 SC 4236).

Part II

JUDICIALIZATION AND THE FRAMEWORK OF REGULATORY GOVERNANCE IN THE INDIAN TELECOM SECTOR

This part examines the concept of judicialization and how in the Indian telecom sector, as elsewhere, judicial processes have come to dominate the regulatory space. It traces the evolution and legislative journey of the Telecom Regulatory Authority of India (TRAI) Act, mediated by judicial oversight. The reasons for industry distress are then discussed, along with the policy response in the initial years. The calling party pays regime and the redefinition of the regulator's ambit in the wake of connected litigation are explored further in the lead-up to the analysis of the amendment of the Act in 2000. The part then proceeds to examine the introduction of the specialized tribunal into the regulatory institutional mix and discusses the interplay between TRAI and the Telecom Disputes Settlement and Appellate Tribunal (TDSAT). Finally, the intervention of the Supreme Court in establishing a judicialized regulatory architecture is examined through the example of the adjusted gross revenue (AGR) cases.

2

TRAI

Origin and early years

2.1 Judicialization and non-judicial legality

It is now fairly common practice in the area of infrastructure to rely on law-backed specialized agencies possessing technical expertise that are insulated from quotidian political pressures to make economically efficient regulatory decisions. The establishment of the Telecom Regulatory Authority of India (TRAI) in 1997 can be seen as one instance of the shift in mode of governance from a hierarchical command and control model to a more 'decentred' version whereby exclusive central ordering by the state is sought to be replaced not by exclusive appeal to the market alone, but by a somewhat more nuanced negotiation rationality that encourages give and take between interdependent actors (Sørensen and Torfing 2004) and that draws attention to a host of economic, political, and legal considerations: the interplay of group interests in shaping the regulatory space, the constrained maximization nature of the regulatory problem, the nature of principal-agent relationship inherent in regulation, and the importance of viewing regulation as long-term relational (and incomplete) contract, as well as the possibilities and limitations of any dependence on self-regulatory instincts. The setting up of TRAI can also be seen as one more instance of rearranging the shifting terrain of separation of powers, since that doctrine, if strictly construed, would not permit rule-making, implementation, and adjudicatory functions to be simultaneously vested in any single pre-existing executive or judicial organizational template. The relocation of legislative power in an executive administrative agency such as TRAI that at the same time sought to be clothed with adjudicatory power is quite different from the classical republican notion of rule by representative majority.

The institutionalization of independent regulation is also a form of judicialization of governance not limited to the direct involvement of judges in the making of public policy and also extends to contexts where decisions are "made in the shadow of judicial processes" (Ginsburg 2008: 6) and where arm's-length transactions are privileged over relational governance. The setting up of TRAI during the period when the Indian government was actively seeking private sector investment in the telecom sector can be seen institutionally as establishing a mechanism Stone Sweet labels 'triadic dispute resolution' (TDR), which involves two contracting parties (the dyad) and a dispute resolver. While the proximate reasons for the emergence of an independent regulator in the form of TRAI have been documented well enough (Athreya 1996; Petrazzini 1996; Sinha 1996; Dokeniya 1999; J. P. Singh 2000; Gupta 2002; Mukherji 2006; Uppal et al. 2006; Kathuria 2007; Raghavan 2007), it is necessary to note that at the broad institutional level, the emergence of TRAI is ultimately a response to the need for a third party to ensure the durability of what were hitherto dyadic relationships of reciprocity but have lately encountered difficulties in sustaining social exchange at an optimal level[1] and to regulate the behaviour of participating entities in accordance with rules. Obviously, there are at least two sets of dyadic forms mediated by TRAI: the first is the duo of telecom licensor (embedded in the hierarchical Indian state) and the largely private sector licensee (steeped in the procedural rationality of the market); the second is the twosome of telecom service provider (operating as we have already noted within market logic) and the consumer of telecom services (who aggregates to almost the entire voting population and can roughly be credited with possessing some kind of normative rationality).

The question of why a sector-specific independent regulatory authority (IRA) becomes institutionally necessary when TDR mechanisms in the form of the judiciary are available to the disputants can be answered by referring to what has been termed 'non-judicial legality', which occurs at the intersection of two trends: (1) judicialization – which aims to produce judicial regimes able to secure credible and predictable property rights – and (2) non-judicial review – which places a premium on a disinterested and technical neutrality brought to bear on economic policymaking by experts – within the modern regulatory space (Morgan 2003). This non-judicial legality offers the social logic of the reflexive "regulation of the process of regulation itself" – a kind of meta-regulation that instals arm's-length regulatory mechanisms systematically into routine policymaking. The answer to the other question regarding why telecom regulation was not left to other non-judicial review mechanisms operating within the same economic rationality (such as competition or antitrust regulatory bodies) can be traced to the distinctive techno-economic features of telecom as a network industry that call for particular forms of expertise in areas such as externalities, essential facilities,[2] interconnection, unbundling, and the like, along with a significant technology component.

That brings us to the question of the timing of this non-judicial legality in the Indian telecom sector – why it occurred at the time it did. The process of instituting this form of legality in the Indian telecom sector has both a legislative history and a judicial one. The coming into force on January 25, 1997, of the TRAI Act, 1997 (preceded by a presidential ordinance promulgated on January 27, 1996), marks India's acknowledgement of and acquiescence to the supranational consensus around ending monopoly national telecom regimes and instituting competition-based markets in telecom services achieved at the conclusion of extended negotiations on basic telecom services at the World Trade Organization (WTO) in February 1997. In this sense, independent regulation as an institutional accompaniment to the liberalization and privatization of telecom – as in the case of other public utilities – was an instance of isomorphism very well occasioned by the composite nature of the principals composed of multiple actors with influences flowing along different dimensions. This first attempt by the government to invest institutional capacity and legal authority in the telecom regulator was not an unequivocal success, and a second legislative effort, through an amending Act in 2000, was needed to concretize the internal logic of independent regulation. The legislative history was influenced not only by factors endemic to the sector and the colour of the prevailing policy consensus in the executive-legislature complex at the central government level, but also by the actions of the higher judiciary, a crucial component of the national regulatory regime. While it was the legislature (in practice invariably led by the executive) that initiated the deregulatory agenda, it was judicial review that effectively controlled crucial aspects of its implementation. Given that the establishment of an IRA would further weaken the doctrine of separation of powers by vesting executive, legislative, and judicial powers in the same institution, the intervention of the courts at critical moments in the regulatory journey may be seen as an attempt to balance the benefits likely to accrue from improvements in sector governance with the costs associated with unitary institutions that tend to reduce accountability.

The judicial history of telecom regulation itself can be described in terms of two fairly distinct periods: initially as a legalistic account of the involvement of courts in order to first institute and then streamline the statutory framework and later on as a medley of interventions that – while sometimes supportive and other times corrective and therefore apparently not entirely coherent taken together – nevertheless can be analytically framed within the judicialization narrative, where such judicialization is seen not simply as "a constitutive framework for market exchange, but also as the arbiter of the balance between competing values struck by economic policies" (Morgan 2003: 494). While in the preliminary constitution of the new regulatory framework for telecom the judiciary appears to have privileged a sharper focus on the legality of specific regulatory arrangements (the rules of the game), once the statutory frame was established to satisfaction, the courts played a more active role in implementation (how the game came to be played). In this second stage even more than in the first, the courts intervened in the regulatory system with a very particular vision of how economic activity needed to be governed in the sector, a vision attuned to the presiding pluralistic economic philosophy. Along the way, the courts came to reassert their pre-eminent role in the dispute resolution process without leaving the battle in the hands of the IRA alone. By shepherding the institution of a statute-backed TRAI, the courts participated in the reduction of transaction costs for telecom sector actors in the first stage; by strictly construing the statute thereafter and reining in TRAI by not allowing it to run away with its agenda, the courts ensured that the realm of making the rules of the game remained with the majoritarian legislature and that the judiciary only retained the power of judicial review. The legislation in the sector hence evolved in a reactive fashion, in spite of a seeming elite consensus on deregulation occasioned by the neoliberal turn in the early 1990s. At the same time, the higher judiciary also oversaw the embedding of a judicial mentality within the regulatory institutional structure by actively encouraging the creation of a specialized tribunal – the Telecom Dispute Settlement and Appellate Tribunal (TDSAT) – and ensuring it operated within the norms of judicial decision-making by balancing its enthusiasm to invest in purely expertise-based adjudication of disputes with caveats emanating from a history of adherence to principles governing the rule of law.

2.2 Origins of the statute

The first legislative attempt to institute a telecom sector IRA referred not only to the National Telecom Policy, 1994 (NTP 1994), vision of providing world class and wide-ranging telecom services to the Indian consumer at reasonable prices by encouraging fair competition among companies registered in India, but also to the need to separate regulatory functions from service providing functions in keeping with the general trend in the world.[3] In seeking to amend clause 6 of section 3 of the Indian Telegraph Act (ITA), 1885, to designate any statutory or non-statutory authority as 'telegraph authority'[4] so the non-statutory IRA, to be called Telecom Regulatory Authority of India, could be given powers related to certain – at that date undefined – assigned functional areas, this was a half-hearted attempt to establish an administrative agency that could be delegated with some of the powers vested by statute in the government. The recognition of the need for a third-party watchdog can be traced to the vaguely worded commitment in NTP 1994 that "suitable arrangements will have to be made (a) protect and promote the interests of the consumers and (b) ensure fair competition" in order to implement the national policy that did not otherwise explicitly mention the institution of IRA.

NTP 1994, while carrying forward the July 1992 opening up of eight services classed as 'value-added services' (VAS) by granting non-exclusive licenses, also sought to encapsulate principles such as company track record, compatibility and usefulness of technology, protection of national security interests, quality and affordability of consumer service, and attractiveness of commercial terms to government that would be used in the selection through tendering of private companies to operate two of these – cellular mobile telephone service (CMTS) and radio paging service – where the number of service providers that could be licensed was constrained.[5] Such an encapsulation was necessitated by the judgment in one particular court case that centred on the selection of CMTS operators in the four metropolitan cities of Delhi, Bombay, Calcutta, and Madras that can be seen as a good example of policymaking under the shadow of judicial decision-making and therefore as an important stage in the judicialization of regulatory governance in the sector. This case had the effect of forcing the executive's hand and pushing the trajectory of regulatory governance in one particular direction.

The Tata Cellular *case and NTP 1994*

The tendering process adopted in the selection of non-exclusive CMTS licensees in the four metros was the subject matter of adjudication in the Supreme Court in *Tata Cellular* v. *Union of India*, AIR 1996 SC 11. Judgment of the three-judge bench (including Chief Justice M N Venkatachaliah) was pronounced by Justice S Mohan on July 26, 1994. The dispute centred around allegations levelled by some bidders of (1) bias, (2) invoking hidden criteria, (3) irrelevant considerations,

(4) bypassing the selection committee, (5) selecting unqualified parties, and (6) faulty marking system by (a second) committee that vitiated the tender. A judgment of the Delhi High Court had the effect of changing the selection, and *Tata Cellular* was the case in appeal against that judgment before the highest court. The Supreme Court dealt with five aspects that emerged for its consideration after arguments concluded: (1) the scope of judicial review; (2) whether the selection was vitiated by arbitrariness (regarding financial projections and rental calculation); (3) whether a key senior official was motivated by bias; (4) whether an apex committee was bypassed; and (5) whether evolving hidden criteria was valid.

The fairly straightforward conclusion of Justice Mohan with respect to the scope of judicial review was arrived at after ranging widely over Indian and international case law. He deduced the following principles from his survey: (a) the modern trend is judicial restraint in administrative action; (b) the court does not sit as court of appeal but merely reviews the manner in which the decision was made; (c) the court does not have the expertise to correct the administrative decision, and if a review of the administrative decision is permitted, it will be substituting its own decision without the necessary expertise which itself may be fallible; (d) the terms of the invitation to tender cannot be open to judicial scrutiny because the invitation to tender is in the realm of contract; (e) the government must have the freedom of contract and that a fair play in the joints is a necessary concomitant for an administrative body functioning in an administrative or quasi-administrative sphere subject however to reasonableness and being free of arbitrariness, bias, and *malafides*; and (f) quashing decisions may impose heavy administrative burden and lead to increased and unbudgeted expenditure. On these principles, the judgment ruled that the selection was not vitiated by arbitrariness, that the allegation of bias did not stand, and that there was no merit in the contention regarding bypassing the apex committee. Most important, the judgment delved into the nature of criteria that were used at different stages to select or reject different bidders and held that the cases of two bidders must be revisited on the basis that these were arbitrary and violated the principles of natural justice. However, other selections were not disturbed, keeping in view the doctrine of judicial restraint.

Tata Cellular was, however, only the beginning. To return to our legislative history, it expedited a kind of policymaking effort and significantly influenced the shape it took. The telecom policy statement also sought to legitimize the actions of the executive branch that had pursued the new economic policy (NEP) line in the telecom sector as well since as early as July 1992 without explicit parliamentary approval and these actions needed to be ring-fenced to stand up to judicial logic – as, for example, revealed in the July 1994 *Tata Cellular* judgment. It is in this spirit that the half-hearted attempt in May 1995 at legislating into existence a non-statutory IRA for the telecom sector by amending ITA should be seen. The uncertain nature of technological development and – paradoxically enough given the deference exhibited by the courts to expert decision-making – the lack of surefooted expertise in the techno-economics of the sector made administrative decisions vulnerable to litigation and, worse still, to being overturned by courts. At the same time, the long-term and incomplete contractual nature of long-gestation utility provision demanded the application of an economic rationality that balanced consumer interests with those of producers. Judicialization doubly reinforced the importance of putting in place an expert body that would mediate the technically complex contracting scenario. The vaguely worded commitment in NTP 1994 to make suitable arrangements in this direction then perhaps reflects executive reluctance to part with potential rent-seeking opportunities occasioned by privatization. That this is not mere speculation is evident from the way the executive continued implementation of the policy objective of inviting private participation in the sector in jurisdictions other than the four metropolitan cities without any change in the institutional arrangements and without constituting an independent regulator. Tenders for grant of non-exclusive (but limited number of) licenses for CMTS and basic service in each of 20 territorial circles[6] were invited by the government in January 1995.

The tension between two divergent conceptions of economic rationality found expression in the turn of events in the legislature when the bill to amend ITA came up for consideration in Parliament in August 1995. The purpose of the bill was to amend section 3 (6) of ITA from its existing form that read:

> "Telegraph authority" means the Director General of Posts and Telegraphs, and includes any officer empowered by him to perform all or any of the functions of the telegraph authority under this Act;

to read as:

> "Telegraph authority" means the Director General of Telecommunication and includes any statutory or non-statutory body or any officer empowered by the Central Government to

perform all or any of the functions of the telegraph authority under this Act;

The Statement of Objects and Reasons indicated that the amendment was proposed to enable the creation of an IRA as a non-statutory body to begin with. The bill was debated in the Lok Sabha on August 9, 1995, and August 16, 1995. The stated intention of the government was to gain some experience of independent regulation by first constituting a non-statutory multi-member body headed by a serving or retired Supreme Court judge or a serving or retired chief justice of a High Court in order to make good on the commitment already made to attract investment and the policy stipulation to protect consumer interest and ensure fair competition and, when the time came, to convert it into a statutory body. Assurance was provided that the proposed IRA would be autonomous, with clearly defined functions, responsibilities, and powers, although these were not spelt out in any kind of detail. The bill was passed by the Lok Sabha on August 16, 1995. However, on the date of its listing in the Rajya Sabha on August 22, 1995, the bill was not moved and was not taken up for consideration.

The exact reason why the government that steered the bill in the Lok Sabha and ensured its passing on August 16, 1995, agreed to a reconsideration when its turn before the Rajya Sabha came up on August 22, 1995, is not entirely obvious from the discussions in the House on that day. The process of privatization continued its course, and the evaluation of tenders for CMTS basic service licenses in the department of telecommunication (DoT) of the government went on.[7] The bitterness that accompanied the debate over the proposed withdrawal of the Indian Telegraph (Amendment) Bill in the Rajya Sabha was based to a large extent on differences of opinion over the way the tender process was being conducted by the executive branch, with the government and the opposition blaming each other for not instituting an IRA in spite of general agreement as to its necessity to oversee privatization (PRS 1995b). The setting-up of TRAI continued to be stalled, in spite of specific commitments made in the guidelines and tender documents. Under the circumstances, it was once again left to the Supreme Court to provide the catalyzing impetus for ensuring the IRA was set up and the sector-specific TDR mechanism was established by statute via its judgment in the *Delhi Science Forum* case.

The Delhi Science Forum *case*

As in *Tata Cellular*, it was the tendering process for grant of licenses that came under challenge in *Delhi Science Forum and Ors.* v. *Union of India and Anr.*, AIR 1996 SC 1356. If the metro CMTS tenders were at issue in the earlier case, the challenge in *Delhi Science Forum* was to the power of the central government to grant licenses to different non-government companies to establish and maintain telecom systems in the country and the validity of the procedure adopted for the purpose, in the case of *both* CMTS and basic service licenses in territorial circles even though the judgment explicitly mentions only basic service licenses.[8] Writ petitions were filed before the Supreme Court, and the court also transferred some petitions filed in other High Courts. Judgment in the case was pronounced by Justice N P Singh, from a three-judge bench (which once again included then chief justice A M Ahmadi) on February 19, 1996.

The judgment referred to India's liberalized economic policy as a reason for seeking acceleration in the sector. It noted that the challenge was to the validity of the new policy on the grounds of (1) endangering national security and (2) not serving national economic interest. The court also noted that since NTP 1994 was placed before Parliament (and therefore deemed to have been approved by it), the scope of judicial review was limited to whether any legal or constitutional bar existed on its adoption and no further.[9]

The non-creation of IRA was one of the grounds of challenge, and the observations of the court reflect the far-reaching impact of judicial thinking on the legislative history of the TRAI Act. The court noted international experience (specifically the UK, the US, Australia, France, Japan, and generally in many other developed and under-developed countries) and also the fact that TRAI Ordinance 1996 was promulgated after the hearing of the writ petitions concluded.[10] The court went on to emphasize that the existence of a TRAI with appropriate powers was essential for the introduction of plurality in the telecom sector so as to harness individual appetite for private gains for social ends. It took note of the provisions of the ordinance and observed that the functioning and activities of different telecom service providers could be regulated accordingly. The writ petitions were dismissed subject to the directions given.

Delhi Science Forum yields rich conclusions when viewed through the judicialization lens. The court's endorsement of the establishment of TRAI may be seen as an instance of meta-regulation whereby it sought to embed and systematize juridified regulatory review mechanisms into everyday governmental policymaking in order to secure a credible environment for property rights and contracts in the sector, at the same time acknowledging that the legal domain has come to encompass a wider range of institutions than only courts and judges – institutions distinguished by a "distinctive and observable texture to a set of social practices, rather than the participation of a specified set of actors and institutions" (Morgan 2003: 496). Viewed in this light, the judgment becomes a crucial ingredient in constructing the structure of expertise-oriented regulatory governance that encourages the stable property and contractual rights so crucial to the larger project of judicialized dispute resolution. It exhibited a pragmatic grasp of "exogenous institutional realities" (Levy and Spiller 1994: 209) and balanced the costs of restrained administrative discretion and the benefits of maintaining regulatory flexibility in a rapidly changing technological environment while instituting TRAI. It would be futile to critique the court for its failure to anticipate incoherent legislation or to precipitate a more coherent one. The blame for poorly conceptualized TRAI legislation that did not meet stated objectives must rest with the executive branch.

The ordinance itself followed a comprehensive bill that was prepared and could not be introduced in Parliament even though notice was given for its introduction in the Lok Sabha on November 23, 1995, and was in turn followed by a replacement bill that was introduced in the Lok Sabha on February 27, 1996. This bill too could not be taken up for consideration and passing, and a second ordinance was issued on March 27, 1996; this too lapsed. No authority was constituted under either ordinance. The third attempt at legislating the TRAI Act was by way of the TRAI Bill, 1996, which was introduced in the Lok Sabha on July 23, 1996. This bill was referred to the Parliamentary Standing Committee on Communications for consideration and report on August 25, 1996. The committee gave its report on November 8, 1996.

After receiving the committee's report in the Lok Sabha on November 21, 1996, and incorporating some changes, Bill no. 47 of 1997 to establish TRAI was introduced in the Lok Sabha on March 15, 1997, and passed by it on March 18, 1997, and by the Rajya Sabha on March 20, 1997. Not all the suggestions of the committee found favour with the executive in piloting the bill before Parliament. Of some significance for our narrative are the following inclusions and exclusions from the committee's recommendations: (1) while the definitions of the terms 'licensee' and 'service provider' were included, they did not follow the committee's formulation; (2) the field of selection of chairperson continued to be from among the judiciary, and the recommendation for a broad-based selection committee was not incorporated; (3) the members were to be drawn from a wider pool, but without any consultation with the chairperson as recommended; (4) the number of members was kept between two and six; (5) the salaries and allowances were to be as prescribed separately; (6) the additional advisory functions recommended by the committee were incorporated into the functions listed in section 11 (1); (7) the recommendation to bring tariff setting under section 11 (2) under Parliamentary oversight was not included; (8) provision for government departments qua service providers within the ambit of section 12 (2) was included; (9) appeal continued to remain to the High Court and not to the Supreme Court as recommended; (10) penalties were made more stringent, but imprisonment was not included; (11) the need for TRAI to obtain the government's prior approval before issuing regulations was excluded; and (12) the maintenance of interconnect agreements was included as a function in section 11 (1) of the Act.

After presidential assent, the bill as passed by Parliament became the TRAI Act, 1997, on March 28, 1997, replacing yet another Presidential Ordinance (No. 11 of 1997) issued on January 25, 1997.[11] TRAI was to be constituted[12] with a chairperson (a serving or retired judge of the Supreme Court or chief justice of a high court) and between two to six members (who possessed special knowledge in specified areas), and they could not be removed from office without making a reference to the Supreme Court. The powers and functions of TRAI were laid down in chapter III of the TRAI Act, 1997. Section 11 (1) of the act, opening with a non obstante clause vis-à-vis "anything contained in the Indian Telegraph Act, 1885", set out 16 functions TRAI was to perform in sub-sections (a) to (q), along with a residual provision in sub-section (r) requiring it to perform such other administrative and financial functions as may be entrusted to it by the central government. Much more unequivocally, and once again opening with the same non obstante clause, section 11 (2) of the Act vested tariff-setting powers in TRAI that may, from time to time, by order, notify the rates at which telecom services shall be provided. Section 11 (3) provided that TRAI shall not act against India's sovereignty, integrity, security of the state, friendly relations with foreign states, public order, decency, or morality, while section 11 (4) required it to ensure transparency while exercising its powers and discharging its functions. Chapter IV of the act, in providing for TRAI to settle disputes relating to (1) technical compatibility and interconnection between service providers, (2) revenue sharing arrangements between them, and (3) quality of service (QoS) and consumer interest (section 14 (2)), elaborated in section 14 (1) that such disputes shall be among service providers or between service providers and a group of consumers and provided that these did not include disputes that were subject to the jurisdiction of the MRTPC,[13] the consumer courts, or those under section 7B (1) of ITA. TRAI was vested with the powers of a civil court for this purpose and was to be guided by principles of natural justice. Its orders and directions were to bind service providers, government, and all others concerned. Fines for non-compliance were prescribed, and appeals were to lie with the High Court.

The TRAI Act, 1997, as first enacted, has been widely criticized for the weak and infirm mandate it gave the IRA. The history of the pre-amendment TRAI Act shows the multiple influences affecting its course, including both brazen and nuanced political, economic, and legal justifications. By not incorporating all the suggestions of the Standing Committee, the executive leadership did itself no credit by maintaining ambiguity on crucial aspects of the statute. The grudging acceptance by the executive branch of the need to create TRAI "because privatization without a regulator could not secure the property rights of investors" (Mukherji 2009: 513) led to imprecise role stipulation. It has been argued with hindsight that while the TRAI Act gave wide-ranging powers to the regulator, discrepancies between it and ITA created jurisdictional conflicts between TRAI and the government acting as service provider, thereby exacerbating the problems created by the delay in setting up the IRA that allowed the government to retain "both policymaking and regulatory authority in the process of induction of its competitors"[14] (Dokeniya 1999: 123). Many of these arguments were also articulated in the Parliament during the debates for passing the bill in 1997. Table 2.1 gives the important dates in the evolution of regulatory governance until TRAI was set up and points to the key role played by the higher judiciary at crucial stages in that process.[15]

The Delhi High Court cases of 1998–2000 and NTP 1999

The early litigation before TRAI as adjudicator involved private telecom companies appealing against "predatory conduct by the DoT that sought to drive up prices for cellular service" (Kathuria 2007: 201). The challenges were to executive decisions emanating from DoT

Table 2.1 Chronology until TRAI was established

Date	*Event*	*Remarks*
July 1991	New Economic Policy	Budget speech and Statement of Industrial Policy on July 24, 1991
January 1992	Technical bids invited for metro CMTS	
July 1992	VAS opened to private investment	
February 26, 1993	Delhi High Court judgment on metro tenders	Appealed in *Tata Cellular*
May 13, 1994	NTP 1994 announced	No explicit mention of IRA; laid before Parliament
July 26, 1994	*Tata Cellular* judgment of Supreme Court	
September 1994	Guidelines for private sector entry in basic services issued	Institution of TRAI promised
November 1994	Metro CMTS licenses issued	
January 16, 1995	Bids for CMTS and basic service licenses invited	
May 26, 1995	Indian Telegraph (Amendment) Bill, 1995, prepared	Bill no. 42 of 1995 introduced in Lok Sabha on June 2, 1995
August 9 and 16, 1995	Debate in Lok Sabha	Bill passed by Lok Sabha on August 16, 1995
August 22, 1995	Amendment Bill listed for discussion in Rajya Sabha	Bill not taken up for discussion
November 23/28, 1995	Notice given for introduction of TRAI Bill in Lok Sabha/withdrawal of Amendment Bill in Rajya Sabha	Could not be introduced/ was not withdrawn
December, 1995	Circle CMTS licenses issued	
January 27, 1996	TRAI Ordinance promulgated	After *Delhi Science Forum* arguments concluded and before judgment pronounced
February 19, 1996	*Delhi Science Forum* judgment	

(*Continued*)

Table 2.1 (Continued)

Date	*Event*	*Remarks*
February 27, 1996	Bill to replace TRAI Ordinance introduced in Lok Sabha	Could not be taken up
March 27, 1996	Second TRAI Ordinance promulgated	Lapsed
July 23, 1996	TRAI Bill, 1996, introduced in Lok Sabha	
August 25, 1996	TRAI Bill referred to Parliamentary Standing Committee	
November 8, 1996	Standing Committee report prepared	Presented to Lok Sabha on November 21, 1996
January 25, 1997	Third TRAI Ordinance issued	
February 19, 1997	Notification setting up three-member TRAI issued	
March 3, 1997	Delhi High Court transfers dispute filed before it to TRAI	TRAI Petition no. 1 of 1997
March 15, 1997	TRAI Bill, 1997, introduced in Lok Sabha	
March 18/20, 1997	TRAI Bill, 1997, passed by Parliament	
March 25, 1997	Chairperson and two members assume office	
March 28, 1997	Presidential assent received	TRAI Ordinance replaced by TRAI Act, 1997

Source: Compiled by author.

that blurred the distinction between its roles as licensor and service provider – decisions such as steeply increasing the tariffs for calls made from predominantly incumbent-owned fixed networks to new entrants' cellular networks; offering only a few points of interconnect to connect to the DoT's fixed and long distance network, which was not yet privatized; and delaying mandatory clearances to private networks to start commercial operations without any moratorium on license fee payments. In this particular case, DoT's stakes as service provider clearly outweighed policy imperatives that were in any case contested at this stage; as a service provider, DoT refused to see merit in encouraging competition, and this perception spilled over to its licensing role as well, leading to a non-acceptance of TRAI as triadic dispute resolver.

TRAI nevertheless rallied bravely to the cause of independent regulation. It held against DoT in these disputes and sought to bring in expertise-based decision-making fortified by adherence to formal legal procedure and precedent. These earliest disputes adjudicated by TRAI that were later taken in appeal to the Delhi High Court were:

a) *Petition No. 1 of 1997* (*Aircel Digilink India Ltd. and Other* v. *Union of India and Others*): The dispute involved the tariff to be charged from fixed network subscribers for calls made to cellular subscribers for intra-circle calls, other than in metro category circles.
b) *Petition Nos. 2 and 3 of 1997* (*Fascel Limited and Others* v. *Union of India and Others* and *Modi Korea Telecom Ltd. and Others* v. *Union of India and Others*): Fascel and others, CMTS licensees, challenged DoT's refusal to defer payment of due license fee and encashment of bank guarantees on the grounds that it had not provided needed clearances, approvals, and timely interconnection, which led to a delay in making services operational and hence to inability to meet license fee commitments.

TRAI's position in these cases has been seen as the assertion of regulatory jurisdiction by the IRA and as marking "the beginning of an antagonistic relationship" between TRAI and DoT (Kathuria 2007: 202; Raghavan 2007: 175). That the DoT did not perceive it in a similar light and continued to act in a manner debilitative of the licensor-licensee dyad became obvious in a further series of decisions the DoT took without bothering to inform TRAI, sticking to the letter of the statute rather than to the spirit of independent regulation. This conduct was not only debilitative of the dyad but also undermined the DoT's own position as it revealed in sharp relief not only the importance of having a third-party dispute resolver but also the necessity of incorporating sufficiently strong normative rules in the office of TRAI. In a manner of speaking, by exacerbating the conflict with TRAI, DoT acted to strengthen that fledgling institution in a way it could not have by simply obeying its directions.

The first case taken to the Delhi High Court under the provisions of the TRAI Act, 1997, involved three kinds of disputes brought before TRAI by private service providers in a set of 13 petitions that impugned the government's actions in licensing and policy formulation that were taken without consulting the regulator. The first of these involved the grant of CMTS license on October 10, 1997, to Mahanagar Telephone Nigam Limited (MTNL), which was the corporatized successor to the incumbent DoT in the metropolitan areas of Delhi and Bombay and ran mainly fixed line services until then. While the tenders for CMTS licenses had indicated a duopoly regime, DoT also reserved the right to operate services directly or through a designated public authority. This provision was invoked to grant license to MTNL in the two metropolitan areas without obtaining recommendations from TRAI and without levying an upfront license fee payable by the licensed private service providers. This grant of license was challenged before TRAI. In a separate development, DoT also issued notices to some radio paging service providers threatening revocation of license for non-payment of committed license fees. These service providers challenged this action on the grounds that TRAI's recommendations were not sought before action for revocation was initiated. In a third development, the government announced a policy for internet on September 16, 1997, again without consulting TRAI. Some e-mail operators challenged this action of the central government in announcing the policy without any recommendation having been made by TRAI in this regard.

TRAI dealt with all three issues in a common order on February 18, 1998, in *Petition No. 3 of 1998 and Civil Miscellaneous No. 1 (Bharti Cellular Ltd. and Another* v. *Union of India and Another)*. After hearing arguments, a two-member bench of TRAI (consisting of chairperson and vice-chairperson) observed that there was no conflict of jurisdiction arising from its functions as enumerated in section 11 (1) and section 4 of ITA that provided for the central government or its delegate – the telegraph authority – to grant licenses. Given the existence of section 38 in the TRAI Act, while the licensing powers of the government were not delegated to TRAI, the order held that "it would not be correct to suggest that even the manner of exercise of powers under section 4 of the 1885 Act have remained unchanged as a result of the provisions of [section 11 (1) (a), (b), and (f)]" (TRAI 1998: 42). It went on to garner legislative intent from the fourth report of the Parliamentary Standing Committee that sought incorporation, among others, of these additional functions into the act. It referred to the specific connotation in economic parlance of the term 'need and timing' occurring in section 11 (1) (a) in the context of liberalization and privatization of public utilities and, after adverting to Canadian and UK examples, stated that "in a multi-operator environment, an independent evaluation of the economic needs is a condition precedent for, on the one hand, maintaining investors' confidence, and, on the other, achieving public policy objectives" (ibid: 44). In its order, TRAI rejected the government's claim to having a right to receive license fees as a licensor without discharging its responsibilities as a service provider to provide interconnection. It also rejected the claim that MTNL was an existing service provider and was therefore not within the ambit of TRAI's recommendatory function. Upon such reasoning, TRAI held that (1) it is mandatory for the central government to have before it TRAI's recommendations with regard to matters covered by provisions of section 11 (1) (a), (b), and (f) – that is, with regard to the need and timing for introduction of a new service provider, the terms and conditions of license to a service provider, and revocation of license for non-compliance of terms and conditions of license respectively; (2) because section 11 (1) (a) was not complied with, MTNL's CMTS (and paging service) license was invalid; (3) MTNL was not an existing service provider in this respect; (4) the internet policy formulated and announced without TRAI's recommendation regarding terms and conditions applicable to service providers under section 11 (1) (b) was invalid; and (5) revocation of paging service licenses without obtaining TRAI's recommendations under section 11 (1) (f) was invalid.

This order of TRAI became the subject matter of the litigation in appeal before the Delhi High Court in *Union of India* v. *TRAI* 1998 Indlaw Del 562. Judgment dated July 16, 1998, was given by single judge, Justice Usha Mehra. The judgment noted that two questions were raised in the petition filed against orders of TRAI: (1) powers of the government to grant or revoke license vis-à-vis the extent and ambit of the powers and functions of TRAI and (2) whether MTNL is an existing service provider. The court observed that the amendment of MTNL's license by the government/licensing authority on the request of MTNL permitting it to commence and carry on CMTS business (and paging service business) in Delhi and Mumbai was challenged before TRAI primarily on the grounds that central government could not amend or grant license without the prior recommendation of TRAI. TRAI, while upholding the challenge, held that the central government cannot exercise its licensing powers without the recommendation of TRAI and that, since the licensor had failed to comply with section 11 (1) (a) of the act, its action stood vitiated. The impugned order of TRAI held that recommendation of TRAI was condition precedent for grant or revocation of license or for laying down terms and conditions of license to be issued. The court considered the arguments of the opposite parties on the terms 'government' defined in section 2 (j) and the use of the expression 'recommend' in sections 11 (1) (a), (b), and (f) of the TRAI Act. It referred to the provisions of the Act and to the report of the Parliamentary Standing Committee. On the question of the definition of 'government' and whether this included government as licensor within its ambit, the court held that it was the department of the government providing telecom services covered under the definition and not government as licensor. It supported its declaration by reference to sections 31, 15 (5), 12 (2) (a), 25, and 11 (1) (o) and (r) of the TRAI Act and a harmonious reading of its provisions with those of ITA, specifically section 4 (1). It also held that grant or revocation of license by the licensor does not fall under the ambit of section 14 (2) of the TRAI Act whereby it was empowered to adjudicate disputes.

The court classified the functions of TRAI (as set out primarily in section 11) into five kinds: (1) advisory (section 11 (1) (a), (b), (f), and (o)); (2) regulatory (section 11 (1) (c), (d), (e), (j), (k), (i), (p), and (q), with powers to ensure compliance provided in section 12 (4) and 13); (3) adjudicatory (section 14 (2)); (4) semi-legislative (tariff-setting powers under section 14 (2)); and (5) ancillary. On the question of whether TRAI's recommendations are mandatory for grant or revocation of license, the court held that the recommendations TRAI was entitled to make under section 11 (1) (a), (b), and (f) cannot be called a condition precedent to the power of the central government to grant license. TRAI's functions under these sections are not regulatory powers and are not binding on the licensor, who is not obliged by statute to seek such recommendations before exercising its plenary, quasi-judicial powers under section 4 of ITA. It supported its decision by reference to sections 38, 25, and 14 of the TRAI Act and section 4 of ITA.

On these arguments, the court held that TRAI erred in concluding that the power of the government to grant or amend license was subject to TRAI's recommendations or that these recommendations were mandatory in nature and set aside the impugned order of TRAI. Disputes regarding grant or amendment of license by licensor were not covered under section 14 (2) of the Act according to the judgment. Justice Usha Mehra's order was challenged before a two-judge bench in the Delhi High Court by the radio paging service providers in *Microwave Communication Ltd.* v. *Union of India and Ors.*, 1999 Indlaw Del 11471. Judgment in this appeal was delivered on October 12, 1999, by Justice Arun Kumar from a division bench that also consisted of Justice Manmohan Sarin.

The appeal adjudicated in this case related to the challenge mounted by radio paging operators invoking the provisions of section 11 (1) (e) of the TRAI Act to contend that ensuring compliance of terms and conditions of license under that section would involve TRAI issuing directions to both licensee and licensor, and the appeal contended that the single judge had erred. In deciding the appeal, the court construed the scheme of the Act in the following manner: the functions of TRAI can be placed in two categories: (1) general, as in section 11 of chapter III, and (2) adjudicatory, as in chapter IV, sections 14 to 20. It then classified the section 11 functions into five: (1) recommendatory/advisory (sections 11 (1) (a), (b), (f), and (o)); (2) administrative or supervisory (sections 11 (1) (c), (g), (h), (i), (j), (k), (l), (m), and (r)); (3) regulatory (sections 11 (1) (e), (p), and (g)); (4) dispute resolution (section 11 (1) (n)); and (5) semi-legislative (section 11 (2)).

The court, agreeing with the single judge, held that central government was not bound to seek TRAI's recommendation while exercising the power to grant license. With respect to the second issue, the court noted that the question that arose for consideration in view of the provisions of section 11 (1) (e) that empower TRAI to issue directions to service providers was whether the word 'government' in section 2 (1) (j) was synonymous with service provider or whether it can have a different, context-specific meaning. The court held that when the government is a licensor, it cannot be a service provider. It noted the suggestion of the Parliamentary Standing Committee on the definition of 'service provider'. The court agreed with the single judge and held that TRAI has no power to issue directions to the government as licensor. Appeals were dismissed.

This judgment given by Justice Usha Mehra (and the subsequent one in appeal that upheld it) has been criticized for "making TRAI's role redundant" (Mukherji 2004: 288) and for ignoring the rationale for the creation of TRAI and focusing "on a narrow reading of the statutory language" (Raghavan 2007: 178). The judgment has also been critiqued for its inability to appreciate legislative intent to lift the quality of telecom service to world-class levels and for failing to "recognize that TRAI was a body equipped with professional capability to give expert advice on licensing issues to the Government" (Kathuria 2007: 203). The significant fallout of the judgment was that "TRAI stopped hearing several pending cases" (Raghavan 2007: 179). While it has been argued that there may have been nothing in the pre-amended Act to support the court's classification of functions entrusted to TRAI under section 11 (1) into different categories, it appeared that the logic of the act of classification had an enduring effect on the way the various actors – not excluding TRAI itself – perceived the role of the independent regulator. Certain other developments that had important effects on the regulatory environment also came about as a result of the judgment. It is necessary to recapitulate some of these before proceeding to explicate the regulatory-judiciary interface in the telecom sector.

2.3 Industry distress and policy response

It needs to be mentioned that the nature of the disputes essentially centred on the investment distress experienced by licensees who had obtained licenses for a certain business proposition that was unraveling in the face of policy formulation that had been inadequately thought through and was compounded by inconsistencies in implementation.[16] The inadequacy in the loose framework that housed the long-term contractual relationship between licensor and licensee was further exacerbated by the non-institution of an IRA vested with clear-cut TDR functions. The perceptions of the principal (the central government, through its legislative effort in crafting the TRAI Act, 1997) and of the agent (TRAI as it construed its functions and powers under that act) diverged significantly to the detriment of the licensees' business plans. The disputes raised by licensees revolved around either seeking implementation of DoT's obligations under the license or absolving the licensees from fulfilling their end of the bargain in terms of payment of license fees or providing liquid guarantees. This led to a situation where the licensees defaulted in the payment of license fee dues and took a hit in profitability. It has been observed in this context that "licensees of cellular mobile and basic services did not honour their contractual obligations and DoT also showed laxity towards them in taking action as per license conditions resulting in mounting of licence fee dues to the extent of Rs 3779.45 crore against the licensees as of 31 May 1999" (Comptroller and Auditor General of India [CAG] 2000: Paragraph 14.4). Two study reports, one in June 1998 by Industrial Credit and Investment Corporation of India Limited (ICICI) – an interested party with financial stakes in the sector – and a second in November 1998 by the government's Bureau of Industrial Costs and Prices (BICP), also assessed the state of cellular telecom projects and industry and supported the conclusion that the telecom licensees deserved concessions in order to stay afloat.[17] Following these developments, the central government set up a group on telecom (GoT) on November 20, 1998, headed by the deputy chairman of the Planning Commission, who subsequently went on to become the minister of external affairs (also the minister of electronics), and consisting of 11 other members initially, to make recommendations on a proposed new telecom policy, measures for resolving the issues affecting CMTS and basic service licensees, and other issues pertaining to the functioning of TRAI.[18] As an interim measure, extension of time for payment of license fee dues was also granted to the licensees subject to some conditions.

The macroeconomic environment too had changed in the five years that had elapsed since NTP 1994 was formulated. The government responded with the New Telecom Policy 1999 (NTP 1999), which was approved by the Cabinet on March 26, 1999, took effect from April 1, 1999, and made significant changes to the sector's governance structure. NTP 1999 was born of a stated recognition that the participation of the private sector fell short of expectations as well as the realization that developments in the five years since NTP 1994 required that the licensing framework be reworked to accommodate the effects of convergence of markets and technologies. It foresaw communication infrastructure in terms of eight service provider groups: (1) access providers, including cellular mobile service providers (CMSPs), fixed service providers (FSPs), and cable service providers; (2) radio paging service providers; (3) public mobile radio trunking service providers (PMRTSs); (4) national long distance operators (NLDOs); (5) international long distance operators (ILDOs); (6) other service providers (OSPs) who did not provide switched telephony; (7) global mobile personal communication by satellite (GMPCS) service providers; and (8) VSAT based service providers. NTP 1999 also promised to resolve problems faced by existing licensees in a legally tenable manner consistent with their contractual obligations. For the crucial categories of CMSP and FSP, the policy extended the period of license from the erstwhile 10 and 15 years respectively to a uniform 20 years (extendible by 10 years) and, even more significantly, announced that the licensees would be required to pay a one-time entry fee and license fee as a share of revenue after obtaining TRAI's recommendations on the bases for determining entry fee, selection of additional operators (other than DoT and MTNL, which would enter as the third CMSP operator in each service area), and percentage of revenue share. NTP 1999 also provided for direct interconnection between CMSP and FSP networks and other networks in their respective service areas, with additional provision for entry into NLD services after TRAI's recommendations were obtained. Wireless in local loop (WLL) technology was permitted for faster rollout of FSP networks, again to be based on TRAI recommendations on entry fee and revenue sharing license fee arrangements. NTP 1999 also promised to open up ILD services and corporatize DoT's service provider function.

In committing itself to a strong and independent regulator, NTP 1999 stated that (1) it would be clarified that TRAI had powers to issue directions under section 13 of the TRAI Act to the government in its role as service provider and further that its adjudicatory role under section 14 of the Act would also extend to government acting as service provider; (2) TRAI would be assigned a role in the arbitration of disputes between licensor and licensee; (3) the government would invariably seek TRAI's recommendations on the number and timing of new licenses before issuing them; and (4) the functions of licensor and policy-maker would remain with the government, and where TRAI had a recommendatory role, it would not be statutorily mandatory for the government to seek its recommendations.

However, NTP 1999 was yet only policy; the statute hadn't changed by its date of effect on April 1, 1999.[19] Only section 11 (2) of the act, dealing with the tariff-setting function, appeared to be least controversial in terms of affording autonomous decision-making by TRAI. It was under the auspices of this section that TRAI went ahead to issue the Telecommunication Tariff Order 1999 (TTO 1999) on March 9, 1999 (even as the exercise to change the policy framework was underway within the central government) after going through an extensive consultation process with "all stakeholders in the sector" signalling that (1) "tariffs will be further re-balanced towards costs" and (2) service providers (and, through them, customers) "will be provided enhanced flexibility for pricing" (TRAI 1999b: 52–53), in spite of stated major disagreements on various aspects and implications of the order with DoT. On a parallel track, the consultation process initiated in November 1997 to address the concepts, principles, and methodologies related to telecom tariffs and commercial aspect of interconnection was moved forward by TRAI with the release in September 1998 of specific proposals for interconnection charges and revenue sharing arrangements between service providers and culminated in the issuance of the Telecommunication Interconnection (Charges and Revenue Sharing) Regulation 1999 (TIR 1999) on May 28, 1999, under the powers conferred under sections 11 (1) (c) and (d) of the TRAI Act, 1997 (TRAI 1999c). Together, TTO 1999 and TIR 1999 represented the thrust of the economic expertise TRAI sought to bring to bear on the dynamics of the telecom sector in India. Both these regulatory interventions not only explicitly noted elements of the new policy thinking (the draft policy in the case of TTO and NTP 1999 in the case of TIR) on aspects of the determinations made but also went on to articulate and implement new economic thinking in tariff setting and the regulation of interconnection and revenue sharing arrangements.

Among other proposals that radically altered the way costs were translated into consumer prices in the sector through the provisions contained in these two determinations was one that sought to introduce the calling party pays (CPP) system for cellular mobile telephone subscribers who until then had been paying for incoming calls as well. Introduction of the CPP regime implied an increase in the price of the incumbent's subscribers' fixed line to mobile telephone calls, with a portion of the increased charges accruing to the mobile service provider via mandated revenue sharing arrangements. As the DoT indicated it would "take some time to make the technical adjustments required to implement the CPP system", TTO 1999 proposed CPP would be introduced by August 1999. After a slight delay, TRAI went ahead and unveiled the CPP regime by way of two further regulatory determinations: (1) the first, amending TTO 1999, was the Telecommunication Tariff (Fifth Amendment) Order 1999, and (2) the second, a regulation amending TIR 1999, was the Telecommunication Interconnection (Charges and Revenue Sharing – First Amendment) Regulation 1999. Both were issued on September 17, 1999, to allow the new CPP arrangements to take effect from November 1, 1999. Both were promptly challenged before the Delhi High Court by different parties.

2.4 The CPP case

The challenges to the CPP regime in the Delhi High Court were decided in *Mahanagar Telephone Nigam* v. *TRAI* and *Telecom Watchdog* v. *TRAI, 2000* Indlaw Del 330. Common order in judgment in both petitions was delivered on January 17, 2000, by Chief Justice S N Variava on behalf of a two-judge bench.

The petitions impugned the interconnection regulation and the amendment to the tariff order of TRAI – both dated September 17, 1999 – mainly on the ground that TRAI did not have the power to issue any regulation which affected the rights of individuals under contracts or sought to override terms and conditions of licenses issued by the central government to various parties. While other grounds for challenging the tariffs and charges were also indicated, all parties agreed the issue of TRAI's power needed to be first decided.

In its judgment, the Delhi High Court noted that functions of powers vested by statute in TRAI, regulatory or otherwise, were qua service providers only and that, in all matters concerning the central government not acting as service provider, TRAI had only recommendatory powers. The court divided the functions under section 11 (1) in two: (1) recommendatory (sections 11 (1) (a), (b), (f), (k), and (o)), where it can only make recommendations or give advice, and

(2) non-recommendatory (the rest of the sub-sections except (l) prescribing maintenance of register of interconnect agreements and the residual (r)). The court referred to *Microwave Communication* and stated that this judgment was binding and that the court agreed completely with its reasoning. While the recommendations/advice were to be given to the central government, TRAI did not have any function or power qua the central government in the non-recommendatory functions. The government was not bound to seek recommendation or advice or to follow it if such advice was tendered.

The court observed that what cannot be done directly cannot be done indirectly. It also observed that provisions of a statute have to be constructed harmoniously. In its view, if it was accepted that the non-recommendatory functions could be exercised by TRAI unfettered by its recommendatory functions, there would necessarily be a clash between the two kinds of powers. It referred to examples such as section 11 (1) (c) and (o) and section 11 (1) (b) and (d). Specifically regarding the latter, the court observed that TRAI, purportedly in the exercise of powers under section 11 (1) (d), had overridden the contractual provisions between parties and affected the rights created by the license, thereby converting a recommendatory function into a directory (read mandatory) power. It observed that TRAI could not directly or indirectly vary the terms and conditions of the government in a license to a service provider. When the legislature intends to confer on a body the power to vary contracts or existing private rights, it has to do so specifically, and no such power can be assumed. If the argument (on each set of powers being exercised independently) is accepted, it would amount to conferring on TRAI the power to vary private rights not specifically conferred by statute. Thus, the only manner in which the non-recommendatory powers can be read in harmony with the recommendatory power is to see that the regulatory powers are exercised within the framework of policy and decisions taken by the central government and within the framework of the licenses issued by the central government. The court also noted that a policy decision of the government need not only be conveyed to TRAI under section 25, but can also be ascertained from the terms and conditions laid down by the government in the various licenses. The court did not agree that such a reading would make the recommendatory functions redundant. It noted that these are 'substantial powers' and assured that the central government would give due weight to TRAI's recommendations. The judgment termed the non-recommendatory functions as 'policing functions' and said these were vital and important functions as well.

The court observed that the only provision in the Act that permitted TRAI to issue an order binding even on the central government not acting as a service provider was section 11 (2), given the background of the provision in the licenses that allowed for rates to be varied. Hence this power could be exercised even in the absence of any specific provision in the statute that it could be exercised notwithstanding any contract or decree or order of court. The court also observed that another reason the argument could not be accepted was the adjudicatory power vested in TRAI by virtue of section 14. It held that TRAI could not be allowed to adjudicate disputes arising from or as a result of its own regulations/directions. On section 11 (1) (d), the court observed that even TRAI realized that its function under this section was to intervene only in the event of the service providers not being able to arrive at an arrangement for sharing of revenues.

On these arguments, the court quashed the regulation and order. The court also noted developments that would affect the introduction of the CPP regime and directed that the CPP regime could not be introduced as ordered by TRAI on October 20, 1999, since the new revenue sharing regime mandated by NTP 1999 had been introduced by the time the judgment was delivered. Since the regulation had been struck down, the CPP regime could not be introduced in the form ordered by TRAI. It directed the TRAI to issue necessary fresh directions keeping in mind consumer interest and also whether such a regime could be introduced within the terms and conditions of existing license and within the policy framed by the government.

This judgment in *MTNL* v. *TRAI* rendered in January 2000 confirmed the judicial reckoning of TRAI's statutory mandate first outlined by Justice Usha Mehra in *Union of India* v. *TRAI* in July 1998. That not only the judgments in court cases but also the entire process of adversarial litigation – involving as it does both an iterative sharpening of legal arguments and a reappraisal of the factual matrix to enable moulding of favourable judicial outcomes – exercise influence over executive decision-making is trite since the political executive has much to lose if its policy decisions are quashed by the judiciary whether on the basis of the law or the facts of a case. The executive's performance in litigation is in the public domain and is thereby open to independent appraisal and critique; it also has discernible public effects that compel the articulation of alternative policy imaginations. Such was the case in the first decade of telecom sector liberalization in India as well.

Thus, if the period between the opening of the telecom services sector to the private sector in early 1992 to the *Delhi Science Forum* judgment on February 19, 1996, represented the conversion of an executive authority reluctant to forsake the regulatory high ground into a willing accomplice in the creation of the TRAI statute marking an acquiescence of non-judicial legality, the litigation emanating from TRAI's early adjudicatory excursions after it was constituted in March 1997 to the time of the *MTNL* v. *TRAI* judgment of January 17, 2000, was remarkable for the way judicial interpretation of the provisions of the statute profoundly influenced the amendment of that statute, which had the effect of privileging judicialized forms of dispute resolution in the regulatory space. By repeatedly holding that the TRAI Act, 1997, did not provide room to practice what it preached, the higher courts forced the government's hand to reduce regulatory uncertainty and imprecision. The following chapter demonstrates aspects of this profound impact of judicial thinking on the legislative process.

Notes

1 These difficulties arise from privatization and marketization, which demand more structured rules for coordination and dispute resolution. Provision for licensing telecom services is as old as 1880, when the British Post Office allowed telephone companies to carry on business in London under license whose terms included a validity of 31 years, payment of royalty equal to 10 per cent of gross revenues, restriction of area of operation, and prohibition of use by non-subscribers, with the postmaster general reserving the right to purchase the exchange system on expiry of certain periods of time (Hazlewood 1953: 14). The Indian statute of 1885 too reflected this administrative regime, but it was not until the 1990s – when privatization was instituted and licenses increasingly sought to be issued – that a third party (TRAI) was considered necessary to mediate the dyadic licensor-licensee relationship.

2 While the doctrine of essential facility has antitrust origins, it has different implications in the telecom sector where it is used to mandate sharing of essential facilities with competitors rather than to prevent monopolies from occurring (or breaking them down as they occur).

3 This is more or less the language used in the Statement of Objects and Reasons (dated May 26, 1995) in the Indian Telegraph (Amendment) Bill, 1995 (no. 42 of 1995), which sought to amend section 3 of the Indian Telegraph Act, 1885 (ITA).

4 The term came on the statute book in 1885 when the director general of telegraphs (later posts and telegraphs) was designated as the telegraph authority for the delegation of the colonial government's monopoly powers over establishing, working, and maintaining telegraphs (which was defined in a wide enough manner to include telephones).

5 The constraint is evidently the availability of 'free' electromagnetic spectrum, which is a technical input into these two services. Licenses for the other six services – which were truer to the nomenclature in that they added value to existing telephony services – were available on a

non-exclusive basis. Again, it is interesting that NTP 1994 nowhere discusses spectrum issues.

6 The number was 21 for basic services, where Delhi metropolitan circle was also included. The telecom licensing regime in India is based on the concept of territorial circles that roughly correspond to the jurisdictions of her states, with some variations as in the clubbing of some states in the north east and splitting of Uttar Pradesh state in two. Each of these was classified under 'A', 'B', or 'C' category based on perceptions regarding potential for growth and revenue earning, with category A considered most valuable. This is a legacy of how monopoly service provision has been administratively handled since independence.

7 A healthy number of bids were received by June 1995: 33 bids for CMTS in 18 of the 20 circles on offer and 81 bids in basic services in 20 circles (Uppal et al. 2006: 20).

8 Tenders for both services were invited in January 1995.

9 At one point, the judgment declared NTP 1994 to be a historic departure from the practice followed during the last century.

10 The ordinance was promulgated on January 27, 1996.

11 This third ordinance was acted upon to constitute a three-member TRAI on February 19, 1997, under a notification issued by the central government. See paragraphs 8 and 9 of TRAI's order in *Petition No. 1 of 1997* (TRAI 1998: 2). The Lok Sabha and Rajya Sabha debates on March 18 and 20, 1997, also concerned statutory resolutions to disapprove of the ordinance itself, along with passing the bill replacing that ordinance (LSD XI 1997; PRS 1997).

12 The chairperson, a vice-chairperson, and one member were appointed by a notification dated February 20, 1997, and they assumed charge on March 25, 1997; another member joined on January 5, 1998 (TRAI 1999a: 8). Three other members were to join TRAI in December 1998 and January 1999.

13 The Monopolies and Restrictive Trade Practices Commission was the earlier avatar of the Competition Commission of India.

14 As late as in February 1997, India signed up for a regulatory reference paper that was appended to the WTO agreement on basic telecommunications without committing to an IRA (Uppal et al. 2006). The modified reference paper commitments are summarized in TRAI 1999a: 54–55.

15 The role of political institutions is significant, especially in the way interdepartmental turf battles within the executive branch came to be resolved in the body of the act (see Mukherji 2009). What is particularly pertinent is the profound influence of the judicial pronouncement in moulding the larger process of statute creation.

16 The demand projections of the licensees had also gone awry.

17 It has been observed that on November 1, 1997, the government approved a reference to TRAI to study the financial health of the cellular industry, but DoT did not entrust the job to the regulator but preferred to obtain reports from ICICI (which was processed on DoT file in June 1998) and BICP (which gave its report on November 9, 1998), on the stated ground that TRAI had little institutional capacity at that point to take up the study (Joint Parliamentary Committee [JPC] 2013: 6–12). The conclusion that the ICICI's report "mostly contained recommendations which were

in line with the concessions sought by cellular operators" (ibid: 148) is unsurprising.

18 In a sense, the constitution of GoT could also be seen in the executive's response to the Delhi High Court's judgment in *UoI* v. *TRAI* in July 1998, which brought out the inadequacy of the statutory framework to ensure credible commitments to licensees.

19 NTP 1999 was also placed in the library and secretariat of Parliament on that date.

3

TRIBUNALIZATION AND AFTER

3.1 Amendment of the TRAI Act

Soon after the judgment of Chief Justice Variava was delivered, the government promulgated the Telecom Regulatory Authority of India (TRAI) (Amendment) Ordinance, 2000, on January 24, 2000. The statement of objects and reasons dated February 18, 2000, of the amendment bill (no. 26 of 2000) that was introduced in the Lok Sabha on March 9, 2000, mentions the following salient features of the amendments effected by the ordinance: (1) introducing a clear distinction between the recommendatory/advisory functions and the regulatory functions of TRAI under section 11 (1) of the act; (2) making it mandatory for the central government to seek TRAI's recommendations with respect to the need and timing for introduction of a new service provider and the terms and conditions of license to a service provider; (3) including the functions pertaining to fixing terms and conditions of interconnectivity between service providers, laying down standards of and ensuring quality of service; (4) changing the composition of TRAI to consist of a chairperson and not more than two full-time members and two part-time members; (5) bringing TRAI's tariff-setting functions under the purview of section 11 (3), which prohibited it from acting against the sovereignty and integrity of India, security of the state, friendly relations with foreign states, public order, decency, or morality; (6) making TRAI's regulatory, tariff-setting, and direction-issuing functions appealable to a specialized tribunal and keeping them out of the purview of audit by the comptroller and auditor general (CAG); and (7) introducing provisions for establishing the Telecom Disputes Settlement and Appellate Tribunal (TDSAT) for adjudicating disputes between licensor and licensee, between two or more service providers, and between a service provider and a group of consumers, as well as to hear and decide appeals from any direction, decision, or order of TRAI. Even a bare reading of these objectives and reasons indicates their debt to the judicial oversight of the statute book in the early years of the regulator's existence.

However, in the official narrative, the promulgation of the ordinance was the culmination of a deliberative process by which the central government had constituted a group on telecom (GoT) and information technology (IT) convergence (GoT-IT) headed by the finance minister and constituted on December 13, 1999. The stated reason for promulgating an ordinance instead of passing an Act was the urgency to remove the impediments in the growth of the sector. There was no explicit mention of the court judgments, even though 'certain issues' that were 'hindering the growth of telecom sector' found mention in the statement of objects and reasons of the amending bill. It was left to the parliamentary process to supply this significant omission during the debates in each house that followed, as well as in the report submitted by the Standing Committee on communications in an ex post scrutiny of the provisions of the amending Act in November 2000 (Standing Committee on Information Technology [SCIT] 99–00 2000). The Telecom Regulatory Authority of India (Amendment) Bill 2000 was passed by the Lok Sabha on March 13, 2000, and the Rajya Sabha on March 15, 2000, and received the president's assent on March 25, 2000, to become Act no. 2 of 2000.

The amendment ordinance had one immediate impact: the chairperson and members of TRAI (who by then numbered six, including the vice-chairperson) were summarily removed from office.[1] Only one – an ex–Department of Telecommunications (DoT) bureaucrat – found a place as a member in the reconstituted TRAI that came to be headed by a retired public sector banker. TDSAT also came into existence with the designation of Justice Suhas C Sen, retired Supreme Court judge, as chairperson on May 29, 2000. Two other members of TDSAT came to be appointed in late September and early October 2000. TDSAT started hearing cases from January 2001.[2] In the next section, we briefly sketch the salient features of the statutory provisions that frame the working of this specialized tribunal that – even as it belongs to the 'tribunalization' tradition that has emerged as a distinct feature of the Indian legal system – has nevertheless been described as "a unique quasi-judicial institution, not only to the Indian judicial system, but also to the realm of global administrative law and comparative communication law" (Raghavan 2007: 225). That such a tribunal came to be separated from the independent regulatory authority (IRA) in itself is a marker of the later judicial tendency to strengthen the principle of separation of powers so that "the judiciary and quasi-judicial agencies remain free from executive interference" (ibid: 228). It also marks a shift in the higher judiciary's focus from the substantive task of aiding the conception and constitution of a third-party dispute resolver to the more procedural one of investing the institution with processual legitimacy to withstand judicial scrutiny.

3.2 The specialized tribunal

Specialized courts have evolved as a response to the classical problem of reining in administrative discretion to keep it within the bounds of legislative delegation. While the separation of adjudicatory functions from IRAs and vesting these in sector- or function-specific appellate tribunals has gained common currency in India, the institution of specialized courts and tribunals is not unique to this country. They have a chequered history: for example, in the US legal system where they struck roots in the late nineteenth and early twentieth centuries as administrative tribunals that were sought to be distinguished from administrative agencies in general by the union of administrative, legislative, and/or judicial powers that were seen as required to ensure effective state control of rapidly proliferating private business.

The judicial tendency to privilege the creation of specialized tribunals has kept pace with the reinvention and proliferation of new IRAs. In a subsequent case in the antitrust area (*Brahm Dutt* v. *Union of India*, AIR 2005 SC 730), the Supreme Court noted with regard to a separation-of-powers-based challenge to the appointment of a bureaucrat to head the apex competition commission that the executive took the stand during the course of litigation that an appellate authority would be constituted which would essentially be a judicial body conforming to the concept of separation of powers as recognized by the Supreme Court. It went on to observe:

> if an expert body is to be created as submitted on behalf of the Union of India consistent with what is said to be the international practice, it might be appropriate for the respondents to consider the creation of two separate bodies, one with expertise that is advisory and regulatory and the other adjudicatory. This followed up by an appellate body as contemplated by the proposed amendment, can go a long way, in meeting the challenge sought to be raised in this Writ Petition based on the doctrine of separation of powers recognized by the Constitution.

The interleaving of judicial and legislative processes that led to the creation of TDSAT by amendment of the TRAI Act can be seen as the triumph of the juridical – rather than the administrative or legislative – view of state ordering of private and contractual rights. While undertaking this ordering, the judicial branch did not exhibit the same deference to executive expertise as shown by the legislature, which sought to create an IRA at arm's-length from the overtly political environment. The higher courts, even as they recognized and deferred to the economic rationality that animated such a move towards independent regulation during the early stages, were later reluctant to limit their supervisory role over the administrative agencies to a classical notion of judicial review (whatever the criticisms of that doctrine) and insisted upon writing into the statute a judicialized oversight process in the form of specialized tribunals such as TDSAT that ensured institutional primacy to court-made law and to traditional doctrines of formal justice. No doubt TDSAT (and other specialized tribunals) have been imagined as expert bodies by virtue of statutory stipulations regarding appointments to be made to them. Nevertheless, by first enunciating a separation of the regulator's adjudicatory functions from its other (recommendatory and directory) functions, and then hard-coding the requirement that the tribunals that exercise the adjudicatory functions be headed by judges sourced from the higher judiciary (recall *Brahm Dutt*), the courts have ensured that the tribunals are supervised according to a judicial mentality that seeks to apply general principles of law in preference to domain-specific practices and determinations.

Chapter IV in the amended Act dealing with TDSAT runs through sections 14 to 20, and section 14 indicates two functional areas for TDSAT: (1) under section 14 (a) for the adjudication of any dispute between a licensor and a licensee; between two or more service providers; and between a service provider and a group of consumers, provided that these disputes do not pertain to those under the ambit of MRTPC, consumer courts, or section 7B of the Indian Telegraph Act (ITA); and (2) under section 14 (b) to hear and dispose of appeal against any direction, decision, or order of TRAI under the act. Under section 14B of the Act, TDSAT shall consist of a chairperson and not more than two members to be appointed by notification by the central government; section 14B (2) provides for consultation by the central government with the chief justice of India for selecting the chairperson and members. Section 14C prescribes that the chairperson of TDSAT shall be, or has been, a judge of the Supreme Court or chief justice of a High Court and that the members must have held the post of secretary to the government of India for at least two years or be well versed in technology, telecommunication, industry, commerce, or administration. Section 14D provides for tenure of three years for the chairperson and members, who retire respectively on attaining the ages of 70 and 65 years. The Act also provides for the constitution of benches of the tribunal by the chairperson and the distribution of business among them. TDSAT is not expected by statute to be bound by the procedure laid down by the Code of Civil Procedure, 1908 (CPC), but shall be guided by the principles of natural justice (section 16 (1)). It has been vested with the powers of a civil court (section 16 (2) for CPC purposes and section 16 (3) for the purpose of the Code of Criminal Procedure, 1973 (CrPC)), and every proceeding before it shall be deemed to be a judicial proceeding in terms of meanings and purposes under sections of the Indian Penal Code (IPC). Section 17 stipulates that an applicant or appellant may either appear in person or authorize one or more chartered accountants or company secretaries or cost accountants or legal practitioners or any of its officers to present cases before the tribunal. Appeals against the TDSAT's orders lie to the Supreme Court under section 18, and its orders are executable as decree under section 19.

It would be useful to note that section 15 of the amended Act bars the jurisdiction of civil courts to entertain any suit or proceeding which the tribunal is empowered to determine, and no injunction shall be granted by any court of authority with respect to any action taken pursuant to power vested in the tribunal by the act. This ouster of regular civil court jurisdiction – even though seen as "watertight and complete" – has been seen as not altering the writ or supervisory jurisdiction of high courts under Articles 226 and 227 of the constitution (Raghavan 2007: 235). The Supreme Court's oversight of TDSAT – available under Articles 32 and 136 of the constitution – is made further explicit by provisions of section 18 of the Act.[3]

Initial years of TDSAT and policy developments

TDSAT is vested with original jurisdiction under section 14 (a) and with appellate jurisdiction (over directions, decisions, and orders of TRAI) under section 14 (b) of the TRAI Act. TDSAT started hearing cases in January 2001 under Justice S C Sen and heard a wide variety cases mostly on the original side soon after it started functioning. Table 3.1 gives a summary of TDSAT cases in the first three years of its existence.

In order to more fully appreciate the nature of disputes that came to be adjudicated before the TDSAT, recall that the NTP 1999 framework

Table 3.1 Cases filed before TDSAT (2001–3)

Year	*Case*	*Disposal*	*Chairperson*
Pre-2001	15 petitions transferred from TRAI	January 2001. Summary disposal (mostly withdrawn).	Justice S C Sen
2001	Petition nos. 1, 2, 3 – Limited mobility challenged (Licensor-Licensee dispute)	March 15, 2002, held against petitioners (Cellular Operators Association of India [COAI] and service providers). Later remanded to TDSAT on appeal to Supreme Court.	Justice S C Sen
		August 8, 2003. Limited mobility upheld by majority. Chairman dissented.	Justice D P Wadhwa
	Petition nos. 4 and 5 (L-L dispute)	Dismissed as withdrawn.	Justice S C Sen
	Petition no. 6 – Migration package (Tamil Nadu Circle) (Licensor-Licensee dispute)	March 3, 2004. Held against petitioner service provider with costs.	Justice D P Wadhwa
	Petition no. 7 and 8	May 13, 2004/ February 17, 2004. Dismissed as withdrawn.	Justice D P Wadhwa
	Petition no. 9 (along with petition no. 12 of 2003, no. 3 of 2002, and appeal no. 5 of 2002) – Validity of interconnection regulation (Original side: between service providers; Appellate side: against TRAI decision)	April 27, 2005. Held in favour of private service providers in petitions. Appeal dismissed.	Justice D P Wadhwa
	Petition no. 10 – Interest on license fee (Licensor-Licensee dispute)	April 9, 2002. Held substantially in favour of petitioner licensees. Appeal in Supreme Court upheld TDSAT decision with modification, on March 4, 2003.	Justice S C Sen
	Petition no. 10: Miscellaneous and Execution Applications (along with no. 8 of 2003 and no. 2 of 2004)	March 18, 2005. Interpretation of terms of Supreme Court verdict.	Justice D P Wadhwa

	Petition no. 11 – Denial of points of interconnect (Between service providers)	August 6, 2003. Held against public sector incumbent with costs.	Justice D P Wadhwa
	Petition no. 12 – Charges for leased lines	February 17, 2003. Held against incumbent.	Justice D P Wadhwa
	Petition nos. 13 and 16 – Charges for end links	April 1, 2003. Held against incumbent with costs.	Justice D P Wadhwa
	Petition no. 14 (Between service providers)	August 14, 2001. Dismissed as withdrawn.	Justice S C Sen
	Petition no. 15 – Computation of license fee (Licensor-Licensee dispute)	May 23, 2003. Held against licensee. Appealed in Supreme Court after withdrawing review petition.	Justice D P Wadhwa
	Petition no. 17 – Extension of license effective date (Licensor-Licensee dispute)	February 7, 2002. Held against licensee with costs.	Justice S C Sen
	Petition no. 18 – Limited mobility (L-L dispute) (Leave granted by Madras High Court to approach TDSAT for orders)	August 8, 2001. Interim order. See petition no. 1 of 2001.	Justice S C Sen
	Petition no. 19 (Between service providers)	August 22, 2001. Dismissed as withdrawn.	Justice S C Sen
	Petition no. 20 – Costing dispute (Between service providers) (Along with appeal nos. 7 and 8 of 2001: against TRAI decision)	January 29, 2002. Interim order of disposal pending final determination on principles set out by TRAI.	Justice S C Sen
	Petition no. 21 – Internet licensing policy (Licensor-Licensee dispute)	February 7, 2002. Held against licensee.	Justice S C Sen
	Petition nos. 22 and 23 (L-L disputes)	April 20/May 13, 2004. Dismissed as withdrawn.	Justice D P Wadhwa
	Petition no. 24 – Payments made under license (Licensor-Licensee dispute)	April 9, 2003. Held against licensee with costs.	Justice D P Wadhwa
2002	Petition no. 1 (L-L dispute)	March 13, 2002. Deemed infructuous.	Justice S C Sen
	Petition no. 2 (L-L dispute)	February 2, 2002. Dismissed as withdrawn.	Justice S C Sen
	Petition no. 3	See petition no. 9 of 2001 prior.	
	Petition no. 4 – Incentive scheme for employees of incumbent (Between group of consumers and service provider)	March 8, 2002. Dismissed for lack of jurisdiction.	Justice S C Sen

(*Continued*)

Table 3.1 (Continued)

Year	*Case*	*Disposal*	*Chairperson*
	Petition no. 5 – Limited mobility dispute over use of mobile switching centre by fixed service providers (FSPs) (Licensor-Licensee dispute)	September 27, 2003. Held against cellular mobile service providers (CMSPs) by majority. Chairman dissented.	Justice D P Wadhwa
	Petition no. 6 – Excessive charges by private CMSPs (Between group of consumers and service providers)	July 18, 2002. Dismissed for laches.	
	Petition nos. 8 and 11 – Local area definition (Licensor-Licensee dispute)	December 20, 2002. Held in favour of petitioners. Licensor asked to re-examine contentions.	Justice S C Sen
	Review petition no. 1 of 2003 (in petition no. 11 of 2002)	May 21, 2003. Dismissed.	Justice D P Wadhwa
	Petition nos. 9 and 16 (L-L disputes) and 18 (Between group of consumers and service provider)	March 17, 2003. Disposed of in terms of petition no. 10 of 2001	Justice D P Wadhwa
	EAs in petition nos. 9 and 18	April 27, 2005. No relief granted to petitioners.	Justice D P Wadhwa
	Petition no. 10 – Bank guarantee terms under license (Licensor-Licensee dispute)	February 4, 2004. Held in favour of licensee.	Justice D P Wadhwa
	Petition no. 12 – Earnest money bank guarantee (Licensor-Licensee dispute)	September 12, 2003. Held against licensor, with costs.	Justice D P Wadhwa
	Petition nos. 13 and 14 – Franchisee terms of incumbent public sector service provider (Ordered for TDSAT adjudication by Kerala High Court)	December 19, 2002. Held against Public Call Office franchisees.	Justice S C Sen
	Petition no. 15 (Between group of consumers and service provider)	September 18, 2002. Dismissed as not pursued.	Justice S C Sen

	Petition no. 17 – License fee payment after migration package (Licensor-Licensee dispute)	May 13, 2003. Held in favour of licensees.	Justice D P Wadhwa
	Petition no. 19 – Interest on belated payment after migration package (Licensor-Licensee dispute)	November 14, 2003. Petition of licensee dismissed with costs.	Justice D P Wadhwa
	Petition no. 20 – Reporting requirement for payments under license (Licensor-Licensee dispute)	September 5, 2007. Affirmed determination made by commissioner appointed by TDSAT.	Justice Arun Kumar
2003	Petition nos. 1, 2, and 3 – Interest claim by licensor after migration package (Licensor-Licensee dispute)	April 23, 2003. Held in favour of licensees in terms of petition no. 10 of 2001 as modified by the Supreme Court.	Justice D P Wadhwa
	Petition no. 4 – Interconnection charges (Between service providers)	March 10, 2003. Held in favour of public sector incumbent respondent with costs.	Justice D P Wadhwa
	Petition no. 5 (Between group of consumers and service providers)	January 21, 2003. Dismissed as withdrawn.	Justice D P Wadhwa
	Petition no. 6 – Excess levy of entry fee by licensor on migration challenged (Licensor-Licensee dispute)	February 11, 2005. Held against licensee.	Justice D P Wadhwa
	Petition no. 7 – Definition of adjusted gross revenue (AGR) in license (Licensor-Licensee dispute) (Along with petition nos. 53 of 2004 and 81, 82, and 98 of 2005)	July 7, 2006. Held that there was no effective consultation with TRAI. Remanded for TRAI's recommendations.	Justice N Santosh Hegde
	Petition no. 7 – Definition of AGR in license (Licensor-Licensee dispute) (Along with petition nos. 53 of 2004; 48, 59, 81, 82, and 98 of 2005; and 57 and 102 of 2006)	August 30, 2007. TDSAT considered TRAI's recommendations and made item-wise determination of definition and date of effect. Overturned in appeal by the Supreme Court on October 11, 2011.	Justice Arun Kumar

(*Continued*)

Table 3.1 (Continued)

Year	*Case*	*Disposal*	*Chairperson*
	Review applications nos. 10, 11, 12 in petition no. 7 filed by licensees for date of effect	September 14, 2007. Dismissed.	Justice Arun Kumar
	Petition no. 8 – Effective date for payment of licensee fee after migration package (Licensor-Licensee dispute) (Along with miscellaneous application no. 22 of 2004 in petition no. 10 of 2001, Execution Applications (EA) in petition nos. 16 of 2002, 1, 2, and 3 of 2003, petition no. 3 of 2004, and EAs in petition nos. 9 and 18 of 2002)	March 18, 2005. Interpreted terms of Supreme Court judgment of March 4, 2003.	Justice D P Wadhwa
	Petition no. 9 (Between contracting parties)	April 13, 2005. Dismissed as withdrawn with leave to go for arbitration.	Justice D P Wadhwa
	Petition no. 10 – Implementation of TRAI's determination (Between service providers)	March 29, 2004. Allowed with costs against incumbent service providers.	Justice D P Wadhwa
	Petition no. 11 (L-L dispute)	March 9, 2004. Dismissed as withdrawn.	Justice D P Wadhwa
	Petition no. 12 – Compliance with interconnection regulations (Between service providers) (Along with petition nos. 9 of 2001 and no. 3 of 2002 and appeal no. 5 of 2002)	See petition no. 9 of 2001.	
	Petition nos. 13 to 18 – Tariff for private wire connections (Between group of consumers and service provider) (Ordered for TDSAT adjudication by Kerala High Court)	February 4, 2004. Petitions allowed and increased tariffs revised.	Justice D P Wadhwa

Review petition no. 1 of 2004 in petition nos. 13 to 18 of 2003	July 5, 2004. Dismissed.	Justice D P Wadhwa
Petition no. 19 of 2003 – Payments for interconnection after migration package (Licensor-Licensee dispute)	March 29, 2004. Held against petitioner with costs.	Justice D P Wadhwa
Petition no. 20 – Terms of migration package to paging service providers (Licensor-Licensee dispute)	April 12, 2005. Held against licensees.	Justice D P Wadhwa

Source: Compiled by author.

envisaged a shift to a regime based on revenue sharing (with a one-time entry fee) instead of the unrealistic fixed bids offered by the licensees in their anxiety to corner the licenses. The policy document also provided for extension of license tenure, freedom to obtain licenses for any number of service areas, introduction of additional cellular and basic service providers based on TRAI recommendations, and the entry of DoT/Mahanagar Telephone Nigam Limited (MTNL) as the third CMSP in all service areas. The cushion offered by the guarantee of duopoly was thereby sought to be withdrawn as quid pro quo. The appropriate entry fee and percentage of revenue share were to be recommended by TRAI "in a time-bound manner, keeping in view the objectives of the New Telecom Policy" (NTP 1999).

This process-oriented approach – and the delay it entailed[4] – could not satisfy the pressing demands from the industry for succour. Huge license fee arrears arising from the inability of licensees to pay the promised sums and the consequent litigation were seriously affecting growth in the sector. The prime minister took charge of DoT in August 1999. The vaguely worded guarantee in NTP 1999 was sought to be translated quickly into a tangible plan of action, not least because of intense pressure from the licensees themselves. The government turned to the attorney general for advice on a crucial aspect of the implementation of NTP 1999: whether it was legally permissible to bring incumbent licensees under the new policy regime and in the alternative to advise on the modalities for extending NTP 1999 to circles where existing licensees operated.[5] The attorney general's detailed opinion was received on June 16, 1999.

The government decided on July 6, 1999, in favour of migrating existing licensees to the NTP 1999 licensing regime, "relying largely on the Attorney General's opinion" (Raghavan 2007: 90). The terms of the migration package offered to existing cellular licensees were:[6] (1) the cut-off date for the changeover to NTP fee and license fee as a percentage share of gross revenue will be August 1, 1999; the license fee dues payable until July 31, 1999, were to be treated as entry fee; (3) revenue share was to be paid effective August 1, 1999, while the quantum of revenue share would be decided after obtaining the recommendations of TRAI, 15 per cent of gross revenue[7] of the licensee was fixed as provisional license fee that would adjusted finally subject to the final decision on revenue share percentage and definition of revenue; (4) the licensee would have to pay at least 35 per cent of dues including interest outstanding on July 31, 1999, on or before August 15, 1999, after adjusting prior payment of 20 per cent that was sought in January 1999 to establish licensee bona fides; balance dues were to be paid on or before January 31, 2000, with interest; (5) licensee would have to take action to keep alive, recoup, and enhance bank guarantees to cover outstanding dues by November 30, 1999; (6) both cellular licensees in each service area needed to accept the package, otherwise both would continue under the previous regime for the validity of license tenure; (7) the licensees had to forego the right to operate in a regime of limited competition (duopoly, with DoT/MTNL retaining the right to enter as third operator), and additional licenses without any limit could be issued in any service area; (8) licensee company shareholding was to be subject to 'lock-in' for five years from the effective date of the license with permission to raise additional equity through private placement or public issue; lock-in would not be applicable where shares were transferred for enforcement of financial pledges to creditor banks and financial institutions; (9) the effective date of licenses was notionally extended for the calculation of outstanding licensee fee as an additional relief for all non-metro circle cellular licensees (and basic service licensees); (10) liquidated damages would be payable on existing license terms; and (11) the license period was to be 20 years starting from the effective date of the existing license agreement. The package further stipulated that all conditions needed to accepted in their entirety and "simultaneously all legal proceedings in Courts, Tribunals, Authority or in Arbitration Instituted by the licensee and Associations of Cellular and Basic Service Operators (COAI & ABTO) against DoT or UOI shall be withdrawn". No dispute related to the license agreement for the period up to July 31, 1999, would be raised in the future, and acceptance of the package would be deemed as full and final settlement of all existing disputes. The package also required that amendments to existing license agreements would need to be signed if it was accepted.

The objections to the migration package were elaborately articulated by the CAG in its audit report for the year ending March 1999. Based on the results of a special audit, it argued inter alia that (1) extending the package to basic service licenses (signed only in 1997 and early 1998) that were not covered by the Industrial Credit and Investment Corporation of India Limited (ICICI) and Bureau of Industrial Costs and Prices (BICP) reports was grossly unjustified and premature; (2) the government had already extended benefits to cellular licensees prior to the package based on ICICI and BICP recommendations;[8] since no fresh studies were conducted by DoT, the migration package amounted to a 'double concession'; (3) across the board extension of effective date by six months without individual examination of cases on merit was irregular; granting extension was not recommended by

BICP and was not a point referred to the attorney general, on whose advice the decision was ostensibly based; (4) outstanding license fee arrears could not be treated as entry fees as was expressly mentioned in NTP 1999; this was decided by the government even in the absence of any recommendation by the attorney general; (5) the package was extended prematurely without deciding on key aspects like quantum of revenue share, definition of gross revenue, and prescribing maintenance of records for verification; (6) the license tenure was extended without any linkage with licensee performance; and (7) the question of whether existing licensees were making losses and were eligible for relief was an administrative and financial question that should have been resolved by the government on the basis of the recommendations of the expert GoT, which had not expressed apprehensions on that count; instead the attorney general's advice was used as the basis; even if legally possible, it would create serious financial problems (CAG 2000).

Irrespective of the criticism in Parliament and from the auditor, the migration package found wide acceptance among the licensees. Implementation of the NTP 1999 regime in the form of the migration package was an important juncture at the cusp between the institutional environment (the new institutional economics [NIE] L2 level) and governance (L3) that enabled the licensees to take advantage of primarily first order (but also second order economizing) opportunities. Ironically enough, this was achieved by recourse to freeing the licensee firms from committed contractual obligations rather than through their stricter definition and stringent enforcement. The major changes in the rules of the game occurred due to a manifest crisis occasioned by the ambiguity exhibited by the executive branch in securing the property rights of private investors and the way DoT exploited this ambiguity to engender structural imbalances in the regulatory governance framework. It meant a reduction in transaction costs by realigning external contractual incentive structures for the licensees' businesses that could thereby attract additional investment for network expansion. Given its attractiveness in economic terms, all the cellular and basic service licensees accepted the migration package.[9] DoT too simultaneously requested TRAI for recommendations on the modalities of revenue sharing,[10] signalling further its intention to improve the play of the game. We shall return to the issue of revenue sharing, which became a vexed issue between licensor and licensees and resulted in long drawn out litigation a little later, after briefly examining another early dispute between cellular licensees and DoT. This dispute over what is termed 'limited mobility' witnessed the stiffening of TDSAT's spine by the

Supreme Court, one of the reasons cited by the tribunal when it overstepped its regulatory dispute resolution mandate in its later decision concerning revenue sharing.

Limited mobility litigation

One of the policy initiatives announced by NTP 1999 was to allow fixed service providers (FSP) to build last mile linkages to provide services using wireless in local loop (WLL) technology through radio links to connect subscriber premises with the local exchange. WLL had the potential to change business dynamics for basic service licensees since it reduced their sunk costs significantly. This also presented a problem for fledgling cellular licensees since WLL, while providing last mile access over radio links riding on electromagnetic spectrum, could be utilized to provide subscribers' handheld devices with mobility within the short distance charging area (SDCA) that covered a commercially significant geographical area in many instances.[11] The policy provision regarding permission to use WLL in the last mile was preceded by the announcement by MTNL in early 1999 that it would introduce cellular mobile service using Code Division Multiple Access (CDMA) technology. It launched limited mobility services in Delhi, ignoring TRAI's request to "keep its plans in abeyance while the Authority examined the matter" (Raghavan 2007: 402). Its defiance rode on the decision of DoT in September 1999 that made licenses for cellular service providers 'technology neutral' and permitted use of both Group Speciale Mobile (GSM)- and CDMA-based services. Meanwhile, TRAI's recommendations pursuant to DoT's references of April and July 1999 came to be made on August 31, 2000, and noted inter alia that the business case of the existing FSP licenses was inferior to that of other access service providers (TRAI 2000c). Given the policy endorsement of WLL in NTP 1999 and its contested interpretation with respect to limited mobility services, TRAI was asked by DoT on October 9, 2000, for its recommendations on (1) the scope of the area of handheld subscriber terminals under wireless access system operations of FSPs, (2) the basis for assigning WLL frequency (other than GSM frequencies), and (3) the amount of entry fee and spectrum charges as percentage of revenue to be charged from FSPs (TRAI 2001a). This presented a lucrative business opportunity for the FSPs, whose "businesses had failed to take off in any significant manner" (Uppal et al. 2006: 9), and a setback for CMSPs, as there was little wiggle room for the regulator in view of the approved policy stipulation and the DoT's interpretation as reflected in the request for recommendations. The consultation process that followed saw a sudden and bitter falling out between CMSPs and FSPs – two camps that had "fought aggressively for migration to revenue sharing a few months earlier" (ibid: 8).[12]

TRAI went ahead and made its recommendations on January 8, 2001. It argued that the apprehensions of CMSPs that they would be priced out of the market were exaggerated, that given the limitations on the availability of frequency spectrum the supply of WLL services would be limited, and that WLL with mobility should be permitted subject to it being limited within the SDCA, which would be in the best interests of the consumer (TRAI 2001a). The government was quick to act on these recommendations, and guidelines for fixed services – including limited mobility services – were issued "in a record three weeks, on 25th January, 2001" (Uppal et al. 2006: 9).[13] It has been argued that DoT – supported by federations of chambers of commerce aligned with basic service licensees – was the guiding force for the remarkable "haste with which WLLM was introduced between September 2000 and January 2001" (Mukherji 2009: 509). Among other things, DoT's guidelines of January 2001 allowed applicants for basic service licenses to provide mobility to their subscribers with wireless access systems limited to the local area (the SDCA)[14] in which the subscriber was registered.[15] In an attempt at balancing conflicting interests, it also allowed CMSPs to provide fixed phones based on their GSM network infrastructure. This was not enough to appease the CMSPs, who then "appealed to the legal system for protection of what they considered their legitimate property rights" (ibid: 510). The dispute was litigated before TDSAT and the Supreme Court in three tranches.

Even as the first petition was filed before TDSAT, the guidelines were approved and "as many as 147 applications had been received from 18 companies of whom 14 companies/sister concerns already had cellular licenses" (GoT- IT 2001: 6); the applicants included all but one of the existing FSPs, of which four licensees or their sister concerns were also CMSPs. Letters of intent were issued to 40 applications submitted by three companies. However, in view of the litigation and the bitterness of the dispute, the government decided to refer the issue of limited mobility to the GoT-IT on April 7, 2001, with a stated recognition of the advantages of limited mobility for FSPs for achieving "quick rollout of networks and cheaper service to customers", and asking the group to recommend before April 30, 2001, (1) whether NTP 1999 permitted limited mobility to be offered by FSPs; (2) if permitted, how was it to be introduced consistent with the principle of level playing field between different categories of operators "with the objective of assured service at cheapest possible rates"; and (3) if not permitted, how NTP 1999 could be suitably modified to facilitate limited mobility "to ensure faster achievements of the targets of tele-density as well as rural and remote area telephony at cheaper and affordable rates" (ibid: 16). GoT-IT gave its recommendations on April 26, 2001,[16] stating that in view of "the purposive construction given by the TRAI of NTP-99, permitting WLL technology, coupled with the desirability of encouraging a technology which holds the potential for providing affordable telecom services and increasing tele-density rapidly", it was not necessary to 'reopen' the issue of enabling limited mobility technology to FSPs (ibid: 10). It acknowledged that the more difficult question was how to ensure a level playing field given the divergence of views between FSPs and CMSPs and recommended equalizing the revenue sharing arrangements between FSPs and long distance carriers for WLL subscribers with those between CMSPs and long distance carriers, even as it recognized that "these issues fall squarely within the jurisdiction of TRAI/TDSAT" (ibid: 11). GoT-IT also recommended that the SDCAs be divided into rural, semi-urban, and urban categories and prescribed a formula for FSPs' rollout obligations; it agreed with TRAI's recommendation that no additional entry fee need be charged for the purpose.

With these developments in the background, we now take up the three stages of the limited mobility litigation.

i *First TDSAT Order in Petition no. 1 of 2001*: The petition was filed by COAI and others on January 22, 2001, arraying the Union of India and TRAI as respondents. As the tribunal refused to accede to the petitioners' plea for stopping DoT from taking a decision, the guidelines that came to be issued on January 25, 2001, allowing FSPs to provide CDMA-based WLL services were impugned. Four non-governmental organizations were allowed by TDSAT to become interveners on behalf of 'the general public'. In its full-bench order dated March 15, 2002, the tribunal held that the CMSPs' legitimate expectations were not belied since they too had reaped the benefit of the migration package that allowed them to escape contract obligations and that the migration package in any case envisaged a changeover from limited duopolistic competition to a multiple operator environment. Since new cellular licenses had been issued and the successful applicants "were quite undeterred by the threat of the new WLL technology", and since Bharat Sanchar Nigam Limited (BSNL) and Mahanagar Telephone Nigam Limited (MTNL) – by far the biggest service providers at that point – were to provide WLL phones with limited mobility, cellular licensees could not have a grievance regarding private FSPs who occupied a small segment of the market. The order speculated that the petitioners were attempting to block introduction of the new service that would benefit "common people" only for private gain. It also used the argument that technological progress cannot be blocked by "technicalities and inflexibilities" and that policy must keep pace with technological developments. It did not find any evidence that the government had exercised its authority to change policy disregarding legal procedures or the dictates of reason. The order did not agree with the petitioners' argument that to permit the FSPs to use spectrum without any charge or auction on a 'first come first served' basis was unreasonable since the CMSPs too "have not been charged anything special for use of spectrum". It did not agree that there was no evidence spectrum charges had been built into the fee paid by CMSPs. It noted that there was considerable force in the argument made by the solicitor general that the government "had decided not to make a fortune out of allotment of spectrum to pursue the object of increasing teledensity". It noted that while there might have been erosion in the profits of the CMSPs, they had been compensated in other ways, and that nothing should be allowed to stand in the way of increasing teledensity in the country. On these arguments, the petitions were dismissed by TDSAT. As expected, the dismissal was challenged before the Supreme Court.

ii *Cellular Operators Association of India and Ors.* v. *Union of India and Ors.* (2003) 3 SCC 186: The appeal was heard by a three-judge bench on the Supreme Court that gave its decision allowing the appeals by COAI and others on December 17, 2002. The decision was delivered in two judgments, the principal judgment by Chief Justice G B Pattanaik on behalf of himself and Justice H K Sema, and a concurring judgment by Justice S B Sinha assigning additional reasons. The principal judgment – regarding the purpose and object for which TDSAT was constituted and having examined the provisions of chapter IV of the TRAI Act, particularly section 15, which ousts the jurisdiction of a civil court to entertain any suit of proceeding about any matter TDSAT is empowered by or under the Act to determine – held that the power of the appellate tribunal was quite wide and that "the decisions of the Supreme Court dealing with the power of a court, exercising appellate power or original power, will have no application for limiting the jurisdiction" of TDSAT under the Act. It observed that the stipulation in section 18 (1) meant that "under Section 18 the appeal would lie against any order of the Appellate Tribunal only on the ground specified in Section 100 CPC and necessarily, therefore, it must be a substantial question of law". It held that TDSAT had committed an error by holding that it exercised supervisory jurisdiction and another in holding that its jurisdiction could not be wider than that of the Supreme Court. When its jurisdiction under section 14 was compared with that of the Supreme Court under section 18, it became clear that TDSAT had a much wider jurisdiction than the Supreme Court, which could interfere only on a substantial question of law arising from a judgment of the tribunal and not otherwise. The principal judgment then construed the non-consideration of the relevant material on the issue of level playing field and held that the absence of any finding by TDSAT on that score vitiated its ultimate decision. Such non-consideration on a vital issue was a substantial question of law on account of which the Supreme Court could interfere with the judgment of TDSAT that had concluded baldly that the cellular operators had already been compensated in various ways and also that the entry of the fourth operator had resulted in erosion of profits.

The decision of TDSAT was set aside and the matter was remitted to it "for reconsideration with special emphasis on the question of level playing field, on the basis of materials already on record, after hearing the counsel for the parties concerned". The judges, however, did not enjoin FSPs to stop providing WLL services in view of consumer interest as it was already fait accompli. In his concurring judgment, Justice Sinha noted that TDSAT had arrived at certain findings without application of its mind on vital issues, including the issue of its jurisdiction. On the issue of jurisdiction, Justice Sinha observed that TDSAT had "posed a wrong question and gave a wrong answer". He stressed that TDSAT, by virtue of section 14A (7), was an expert body with wide jurisdiction and that it was entitled to exercise its appellate jurisdiction both on fact and also in law over a decision or order/decision/direction of TRAI. He held that TDSAT's approach – on the premise its jurisdiction was limited or akin to the power of judicial review – was wholly unsustainable. By acting as it did, TDSAT had "failed to take into consideration the amplitude of its jurisdiction and thus misdirected itself in law". Significantly, he held that the "rule as regards deference to expert bodies applies only in respect of a reviewing court and not to an expert tribunal" since an expert tribunal is constituted only with a view to determining the correctness of an order passed by another expert body. The concurring judgment held that the power of the Supreme Court under section 18 was not to be equated with judicial review and that, since it would be concerned with a substantial question of law, its jurisdiction would not be limited to illegality, irrationality, or procedural impropriety in the decision-making process. TDSAT had not formulated any procedure under section 14A (7) and did not go into whether procedural requirements had been followed in accordance with CPC. It was also held that the judgment of a court of tribunal should contain a concise statement of the case, points of decision, reasons for the decisions, and the decisions thereupon. Sufficient or cogent reasons were not assigned in support of TDSAT's findings, and in some cases, no reasons were assigned. On these arguments, Justice Sinha set aside TDSAT's order and remitted the matter to TDSAT for consideration afresh in accordance with law. He did not agree to the respondents' plea that the Supreme Court enter into the merits of the issues at dispute.

iii *Second TDSAT Order in Petition no. 1 of 2001*: On remand, TDSAT's judgment in petition no. 1 (and related petitions) was delivered on August 8, 2003, by which time Justice D P Wadhwa had become chairman. The majority judgment of the two members (who continued from the time of the previous judgment under the chairmanship of Justice Sen) held against the petitioners, with one additional direction to the government to initiate action for levy of entry fee for permitting FSPs to provide limited mobility services, while the chairman, in a minority of one, held substantially in favour of the petitioners and set aside the government's decision contained in the guidelines of January 2001 to permit FSPs to provide limited mobility services.

It is clear from the three judgments discussed here that the result remained the same as far as the introduction of limited mobility services by FSPs was concerned. What is more relevant to the present discussion on judicialization is the liberal definition of the ambit of TDSAT's jurisdiction by the Supreme Court that forced the tribunal to take its mandate more seriously. It has been argued that the cellular operators had a good case on merits and that the tribunal's second order's reasoning "that technological innovations and the need for progress had eviscerated [the distinctions between different classes of telecom services] is unconvincing, slippery, and dubious" (Raghavan 2007: 420). Irrespective of the validity of that contention, it is safe to say from even a bare perusal of the first and second orders of TDSAT that the larger impact of the judgment was in stiffening the spine of TDSAT such that it construed its jurisdiction as a specialized tribunal in wider terms. The clear delimiting of the boundaries of the specialized tribunal with respect to its jurisdiction over TRAI enunciated by the Supreme Court's judgment in the limited mobility case and the ringing endorsement of the amplitude of TDSAT's legal mandate under the TRAI Act as being wider than the judicial review thereof under section 18 established the template of dealings among the government, the regulator, and the tribunal for the future.

The interesting takeaway from the limited mobility episode – especially the "tutorial offered by the Supreme Court to the TDSAT on how it was to envision its own powers and functions more generally" (Thiruvengadam and Joshi 2012: 13) – is its importance to the narrative of judicialization whereby an institution of non-judicial legality (TRAI) was sought to be balanced by the juridification of regulatory governance through investing the oversight body (TDSAT) with a legal mandate of considerable amplitude. The Supreme Court's judgment marked the culmination of the process of creation of the triadic dispute resolver in the Indian telecom sector. It is obvious that each stage in the process – from the very institution of an undifferentiated TRAI to the divestment of the adjudicatory role to TDSAT and finally to the clarification of the scope of TDSAT's adjudicatory mandate – was marshaled by the higher judiciary. Even as expertise came to be privileged as the preferred basis for economic decision-making, the courts ensured that separation of powers was fostered – even strengthened – in the institutions of regulatory governance. By first creating and then imparting a broad remit to TDSAT, the judiciary ensured the supremacy of judicial power in the system of checks and balances that is a concomitant of the doctrine of separation of powers. When an emboldened TDSAT overstepped its limits in construing its enhanced powers, however, the Supreme Court stepped in to remedy matters, as the resolution of the following licensor-licensee dispute illustrates.

3.3 The AGR cases

The migration package offered to the defaulting licensees with effect from August 1, 1999, included a condition that the licensee would be required to pay one-time entry fee and license fee "as a percentage share of gross revenue under the license". While the entry fee was reckoned to be the license fee dues payable until July 31, 1999 (with some adjustment), condition (iii) of the package dealt with the license fee:

(iii) The License fee as a percentage of gross revenue under the license shall be payable w.e.f. 1.8.99. The Government will take a final decision about the quantum of the revenue share to be charged as licence fee after obtaining recommendations of the Telecom Regulatory Authority of India (TRAI). In the meanwhile, Government have decided to fix 15% of the gross revenue of the Licensee as provisional license fee. The gross revenue for this purpose would be the total revenue of the Licensee company excluding the PSTN [public switched telephone network] related call charges paid to DOT/MTNL and service tax collected by the licensee on behalf of the Government from their subscribers. On receipt of TRAI's recommendation and Government's final decision, final adjustment of provisional dues will be effected depending upon the percentage of revenue share and the definition of revenue for this purpose as may be finally decided.

Revenue share percentage was determined in a number of stages for the different services based on the government's consideration of TRAI's recommendations. The provisional license fee of 15 per cent fixed at the time of migration was reduced to 12 per cent for metro and category A circles, 10 per cent for category B circles, and 8 per cent for category C circles with effect from January 25, 2001, consequent upon the issue of guidelines permitting limited mobility. Later on, the revenue share was further reduced by 2 per cent each to 10, 8, and 6 per cent respectively with effect from April 1, 2004.[17] The basis for the levy of percentage revenue share was to be the 'adjusted gross revenue' of the licensee that – once again – was defined not according to the regulator's preferences but in tune with DoT's view of the matter. Table 3.2 indicates the definition of AGR in the case of cellular mobile services and basic services as recommended by TRAI and as defined by DoT in the license agreements and guidelines. It is obvious from the definition adopted that DoT's maximal approach was focused on including all revenues accruing to the licensee, allowing for deductions only on account of actually paid taxes and pass through charges (which in any case would be charged to the gross revenue of the receiving licensee).

The FSPs challenged the definition of AGR before TDSAT in *Petition no. 7 of 2003*. CMSPs – and almost all other service providers whose licenses also contained a definition of AGR largely along

Table 3.2 AGR definitions

Service	*TRAI's recommendation*	*DoT's definition*
Cellular mobile service	'Adjusted Gross Revenue' for the purpose of levying license fee as a percentage of Revenue Share shall mean the 'Gross Revenue' accruing to the Licensee by way of operations of the Cellular Mobile Service mandated under the license (inclusive of revenue on account of value-added services, supplementary services, and sale of handsets) *plus* revenue accruing through resellers, franchises etc., *plus* any revenue foregone through subsidies on handsets or any other rebates, *as reduced* by the following items: (i) Interconnection/ Access charges payable to other Service Providers for carriage of calls; (ii) Roaming revenues collected on behalf of other Cellular Mobile Service Providers and passed on or liable to be passed on to them; (iii) Service Tax paid or payable; (iv) Sale of handsets. It is recommended that the above definition be used for applying revenue share percentage to all category of service areas licensed under the CMS license.	**Gross Revenue:** The Gross Revenue shall be inclusive of installation charges, late fees, sale proceeds of handsets (or any other terminal equipment etc.), revenue on account of interest, dividend, value added services, supplementary services, access or interconnection charges, roaming charges, revenue from permissible sharing of infrastructure and any other miscellaneous revenue, without any set-off for related item of expense, etc. For the purpose of arriving at the 'Adjusted Gross Revenue' the following will be excluded from the Gross Revenue to arrive at the adjusted gross revenue: (i) PSTN related Call charges (access charges) actually paid to BSNL/MTNL or other telecom service providers within India. (ii) Roaming revenues actually passed on to other telecom service providers, and (iii) Service Tax on provision of service and Sales Tax actually paid to the Government, if gross revenue had included the component of Service Tax.

(Continued)

Table 3.2 (Continued)

Service	*TRAI's recommendation*	*DoT's definition*
Basic services	'Adjusted Gross Revenue' for the purpose of levying license fee as a percentage of Revenue Share shall mean the 'Gross Revenue' accruing to the Licensee by way of operations of the Basic Service mandated under the license (inclusive of revenue on account of value-added services, supplementary services, and the sale of handsets) *plus* revenue accruing through resellers (if any), franchisees including CSPs etc. *plus* any revenues foregone through subsidies on handsets or any other rebates, *as reduced* by the following items: (i) Interconnection/ Access charges payable to other Service Providers for carriage of calls; (ii) Roaming revenues collected on behalf of Cellular Mobile Service Providers (if applicable) and passed on or liable to be passed on to them; (iii) Service Tax paid or payable; (iv) Proceeds from sale of handsets.	'ADJUSTED GROSS REVENUE' for the purpose of levying LICENSE FEE as a percentage of revenue share shall mean the Gross Revenue as reduced by: (i) PSTN related call charges (access charges) actually paid to other telecom service providers for carriage of calls; (ii) Service Tax for provision of service and sales tax actually paid to the Government, if gross revenue had included the component of service tax. 'GROSS REVENUE' The Gross Revenue shall include all revenues accruing to the LICENSEE on account of goods supplied, services provided, leasing of infrastructure, use of its resources by others, application fee, installation charges, call charges, late fees, sale proceeds of instruments (or any terminal equipment including accessories), handsets, bandwidth, income from value added services, supplementary services, access or interconnection charges, roaming charges, any lease or rent charges for hiring of infrastructure etc. and any other miscellaneous items including interest, dividend etc. without any set off of related items of expense, etc.

Source: Annexure I, TRAI 2006b.

similar lines – also litigated against the definition. The first TDSAT order common to all similar petitions in the case was pronounced by a two-member bench headed by Chairman Justice N S Hegde on July 7, 2006. The petitioners questioned the validity of the definition of AGR and sought to quash demand notices issued by DoT and refund excess amounts collected as license fee, and – in the case of FSPs – directions to DoT to implement TRAI's AGR-related recommendations given on August 31 and October 31, 2000. The petitioners argued in support of their case that the right conferred on the central government by the proviso to section 4 of ITA to transfer its exclusive privilege of establishing, maintaining, or working of telecommunications for consideration as it thinks fit implied that licenses granted pursuant to such transfer could only relate to the activity of establishing, maintaining, and working of telecom systems. It followed from this implication that the licensor did not have the authority to collect a share of the licensees' revenues derived from other activities. They also contended that in defining AGR, TRAI's recommendations were ignored by DoT in an arbitrary manner without giving them due weight. The petitioners stressed that the language of the migration package clearly indicated the provisional license fee was 15 per cent of gross revenue 'under the license' with effect from August 1, 1999, subject to finalization of the actual percentage after TRAI's recommendations were received.

On the other side, it was argued on behalf of DoT that there was hardly any difference between the definitions of gross revenue indicated in the migration package and the license. In fact, the definition incorporated in the license was more favourable to the licensees than that in the migration package. After having received benefits and monetary gains from the migration package, it was not open to the licensees to question the terms of the license at a later date and wriggle out of their contractual obligations. The licensees could not challenge one condition when the terms of license had been accepted as a package. That the definition did not have an adverse effect was evident from the conduct of the licensees in taking additional licenses after its incorporation in the license agreement. It was also argued that the definition was not contrary to ITA, NTP 1999, or the migration package. While seeking TRAI's recommendations was mandatory, their acceptance was not mandatory under the TRAI Act. TRAI was indeed consulted twice, but after consideration, the government did not find its recommendations acceptable. It was also submitted that DoT had consulted an eminent chartered accountant and that, after considering his opinion as well as the views of CAG communicated in a letter dated December 14, 1999, and after giving due weight to TRAI's recommendations, the definition of AGR came to be incorporated in the license agreement. It was urged that it was not open to TDSAT to interfere with the terms of contract in this petition and or to question the validity or correctness of a condition of the license granted by the licensor.

TDSAT held that the term 'any dispute' was "wide enough to take within its fold all disputes which arise between a licensor and a licensee, which may even be in regard to the terms of the licence". It also held that clause 28.1 of the license agreement specifically nominated TDSAT "for settlement of such disputes which arise out of or connected with the provisions of the agreement". It distinguished the right of the licensor to impose terms and conditions of a license in the telecom sector from the government's rights in an ordinary commercial contract since this right in the telecom sector was controlled by ITA and the telecom policies. It then read the provisions of section 4 of ITA to conclude that if the government chose to appropriate a percentage share of revenue of the licensee as consideration under the section, "then it is logical to conclude that such sharing can be only of gross revenue derived from the transferred privilege of establishing, maintaining and working of telecommunication", and not "as it thinks fit", as argued by respondents. The order did not agree that the petitioners were estopped from challenging the definition of AGR on the argument that the licensees entered into provisional agreements and could not be bound by a clause introduced subsequently. It construed the license fee terms of the migration package in light of the petitioners' arguments that the total gross revenue of the licensee company therein meant total gross revenue of the licensee company *derived from the licensed activity only* and stated that if the intention had been otherwise, the government would have made it explicit at the time of the offer of the package. It held that "neither the argument of estoppels, acquiescence nor subsequent conduct of obtaining additional license without demur can be accepted to reject the contention of the petitioners". The order then turned to the issue of TRAI's recommendations. It once again cited *Delhi Science Forum* and the Supreme Court's observations in the limited mobility case to assert that "the recommendations made by the TRAI have to be given due weightage which the Apex Court expected to be accepted in the normal course which indicates that rejection of such recommendation is an exception and not a rule". It held that DoT's not placing the views of the expert before TRAI was a lacuna that "vitiated the proceedings contemplated under section 11 (1) (a) of the TRAI Act which mandates the Central Government to seek recommendations of the TRAI". It thereby held that a proper and effective consultation with TRAI was not done and that TRAI's views did not receive due weight in the government's decision process.

TDSAT, having come to the conclusion that the mandatory effective consultation with TRAI had not taken place in the process of defining AGR, then adopted the curious course of remanding the matter to TRAI for making recommendations *to the tribunal* within three months. Its order read:

> We think it more appropriate that the matter should be remanded to the TRAI which is the 3rd Respondent herein, before whom the Government should produce the material relied by it while rejecting TRAI's recommendation so that TRAI can consider the same and send its conclusions to this Tribunal and thereafter, this Tribunal will have the benefit of a comprehensive recommendation of the TRAI after considering the materials relied upon by the Government.
>
> Further, while considering the issue now remitted to the TRAI, the TRAI will bear in mind our finding in regard to the inclusion in gross revenue of the licensee revenue derived from non-licensed activities. Apart from that finding, any observation made by us in regard to any particular head of revenue will not be taken by the TRAI as a conclusive opinion. The Authority is free to make its recommendations after independent appraisal of the material that is placed before or obtained by it.

This order was challenged by DoT in the Supreme Court in Civil Appeal no. 84 of 2007 listing as many as 22 grounds for overturning TDSAT's order. By the time the matter came up for hearing on January 19, 2007, in this appeal, TRAI had already made its recommendations in the matter. In view of this, the Supreme Court bench consisting of Justice H K Sema and Justice B Sudershan Reddy held:

> Heard the parties.
>
> Pursuant to the direction of TDSAT in the impugned order, a fresh recommendation has been made by the TRAI. In view thereof, we see no reason to interfere. The appeal is dismissed. The appellant is, however, given liberty to urge the contentions raised in this petition before the TDSAT.

In its recommendations made on September 13, 2006, under the hand of its chairman and another member pursuant to TDSAT's directions of July 7, 2006, TRAI took note of the background of AGR, TDSAT's order, and the submissions made to it by the government and the licensees obtained without going through its usual elaborate consultation process (TRAI 2006b). TRAI's recommendations contain an elaborate discussion of different contested items of revenue and the views of the opposite parties on each of these as well as its own analysis and specific recommendation.

TRAI's recommendations were dealt with in TDSAT's second order on August 30, 2007, by a bench consisting of the new chairman, Justice Arun Kumar, and the same member who had sat with Justice Hegde at the time of the first order. The second order considered each item of revenue and cost on which TRAI had made recommendations and proceeded to make its own determinations after considering the contentions of the petitioners and DoT. Table 3.3 gives a summary of the items and their treatment by TRAI and TDSAT. Thereafter, TDSAT went on to reject the recommendation of TRAI on what it considered to be a stronger legal footing than the contention of the petitioners that it be ignored as gratuitous and beyond the scope of the reference. The order held that once the issue had been finalized by the tribunal, it meant that "at all times gross revenue should have been calculated on this basis". DoT should not be allowed "to reap the benefit of the time lag in completing the legal process". TDSAT ruled that the order would be effective for each petitioner from the date it approached the tribunal.

As could be expected, DoT appealed against this common judgment and order in the Supreme Court under section 18 of the TRAI Act. The licensees, too, filed review petitions in TDSAT seeking the benefit of applying the order from the date of migration to the revenue sharing regime – that is, from August 1, 1999. Upon the rejection of these review petitions on September 14, 2007, the licensees also filed cross appeals before the Supreme Court seeking advancement of the date of effect of the revised definition of AGR. All these cases were adjudicated by the apex court in *Union of India and Anr.* v. *Association of Unified Telecom Service Providers of India and Ors.* (2011) 10 SCC 543. Justice A K Patnaik delivered judgment in the case on October 11, 2011, on behalf of the two-judge bench headed by Justice R V Raveendran.

The Supreme Court formulated four questions of law for adjudication in the case: (1) whether after dismissal of the civil appeal of the government by the Supreme Court's order of January 19, 2007, the Union of India could have re-agitated the question decided in TDSAT's earlier order of July 7, 2006, that AGR would include only revenue

Table 3.3 TRAI and TDSAT views on AGR items

	Revenue/ cost item	*TRAI recommendation*	*TDSAT decision*
1	Income from dividend	Not revenue from licensed activity. Should not be included in AGR.	TRAI recommendation affirmed.
2	Interest income	Only interest calculated on refundable deposits taken from customers to be added to AGR. Interest earned on investment of savings made by licensee after meeting liabilities to be excluded.	Agreed with TRAI recommendation.
3	Capital gains	Revenue on account of sale of immovable property, securities, warrants or debt instruments, and other items of fixed assets should not be included in AGR unless there is verifiable data that the receipts have accrued from establishing, maintaining, or working of telecommunication.	Agreed with TRAI recommendation, without rider. All capital gains to be excluded whether verifiable or not.
4	Gains from forex fluctuations	Should not form part of AGR.	TRAI recommendation affirmed.
5	Reversal of provisions and vendor credits	(a) Revenue arising from reversal of provisions like bad debts and taxes should not form part of AGR. (b) Revenue arising from reversal of vendor credits should be included in AGR.	Both (a) and (b) should be excluded from AGR.
6	Income from property rent	Revenue from property rent should be excluded from AGR provided it is clearly established that the property is nowhere connected to establishing, maintaining, and working of telecommunication.	TRAI recommendation affirmed.
7	Income from rent/ lease of passive infrastructure	Revenue from rent of towers, dark fibre, etc. should be included in AGR.	Agreed with TRAI recommendation.

(*Continued*)

Table 3.3 (Continued)

	Revenue/ cost item	*TRAI recommendation*	*TDSAT decision*
8	Other/miscellaneous income	(a) Revenue streams like sale of tenders, directories, forms, forfeiture of deposits/earnest money, management fees, consultancy fees, and training charges from telecom service should be included in AGR. (b) Revenue from sale of fixed assets in the nature of capital receipts and insurance claims should be excluded from AGR. (c) Payments received on behalf of third party should be included in AGR. (d) Other miscellaneous items and other income to be decided on a case-by-case basis.	Modified: (a) Management consultancy fee and training charges to be excluded. (c) Payments received on behalf of third party should be excluded.
9	Income from sale of equipment including handsets	Not to be included in AGR if discernible and standalone sale of handset or telecom equipment that is not bundled with telecom service.	Agreed with TRAI recommendation.
10	Receipts from Universal Service Obligation (USO) Fund and access deficit charges (ADC)	USO receipts not to be included in AGR. ADC to be included in AGR.	TRAI recommendations affirmed.
11	Inclusion of revenue	Revenue from television uplinking service and internet service (one licensed activity in another) to be included in AGR.	Revenue from these services to be excluded from AGR.
12	Specific items	Costs on account of port charges, interconnection setup charges, leased line, sharing of infrastructure, and roaming signalling charges should not be deducted from AGR.	TRAI recommendation affirmed.
13	Write-offs	Write-off of bad debts or waivers/adjustments/discounts: Bad debts should not be excluded from AGR.	TRAI recommendation set aside. Items under this head to be excluded.
14	Accrual/actual basis	Service tax and interconnection usage charges (IUC) to be shown on accrual basis for both inclusion and exclusion from gross revenue for the purpose of AGR.	TRAI recommendation affirmed.

(DoT did not contest TRAI recommendations on items 6, 8, and 10)

Source: Compiled by author.

arising from licensed activities and not revenues from activities outside the license; (2) whether TRAI and TDSAT have jurisdiction to decide the validity of the terms and conditions of the license (including the definition of AGR) as finalized by the central government and incorporated in the license agreement; (3) whether as a result of the Union of India not filing an appeal against the tribunal's order of July 7, 2006, in some cases that order had not become binding on the Union of India; and (4) whether the licensee could challenge the computation of AGR and, if so, at what stage and on what grounds.

On the first question of law, the judgment looked to the language of the order dated January 19, 2007, in Civil Appeal no. 84 of 2007, particularly the liberty granted to the appellant in that case "to urge the contentions raised in this petition before the TDSAT". The judgment held that under the express liberty granted by the Supreme Court's order of January 19, 2007, each of these grounds could be raised before TDSAT even if it was held that its order dated July 7, 2006, had merged with the Supreme Court's order dated January 19, 2007.

On the second substantive question of whether TRAI and TDSAT had the jurisdiction to decide on the validity of the terms and conditions of the license, the judgment examined the provisions of section 4 (1) of ITA and sections 2 (e), 2 (ea), 2 (k), 11 (1), and 14 (a) (i) of the TRAI Act. It observed that the license granted under the proviso to section 4 (1) of ITA is in the nature of a contract between the central government and the licensee and that as such, "the terms and conditions of the license including the definition of Adjusted Gross Revenue in the license agreement are part of a contract between the licensor and the licensee". It said that while section 11 (1) (a) recommendations have to be given due weight by the central government, these recommendations are not binding on it. It held that TDSAT could not decide a dispute between a licensor and a person who had not become a licensee, with the result that TDSAT had no jurisdiction "to decide upon the validity of the terms and conditions incorporated in the license of a service provider, but it will have jurisdiction to decide 'any' dispute between the licensor and the licensee on the interpretation of the terms and conditions of the license".

The judgment then turned to the condition in the migration package that prescribed the license fee as a percentage of gross revenue, which for this purpose would be the total revenue of the licensee company excluding the PSTN related call charges paid to DoT/MTNL and service tax collected by the licensee on behalf of the government from the subscribers, subject to the government's final decision on both the percentage of revenue share and the definition of revenue after receipt of TRAI's recommendations. It observed that once the licensee had accepted this condition of DoT's letter of July 22, 1999, it could not have approached TDSAT to question the validity of the definition of AGR on the grounds that it cannot include revenues from activities beyond the license since "it was open for the licensees not to undertake activities for which they do not require license under clause (4) of the Telegraph Act and transfer these activities to any other person or firm or company". The inclusion of the definition of AGR in the license agreement was part of the terms regarding payment decided by the central government for parting with its exclusive privilege, and the licensees could not have approached TDSAT for an alteration of the definition of AGR.

On the issue of TRAI's recommendations, the judgment noted the terms of the TDSAT order of July 7, 2006, regarding ineffective consultation with TRAI and observed that TDSAT could have, "at best, if it had the jurisdiction to decide the dispute, directed the TRAI to consider the opinion of the expert on accountancy and send its recommendations to the Central Government and directed the Central Government to consider such fresh recommendations of the TRAI as provided in the provisos to section 11 (1) of the TRAI Act". The judgment faulted the TDSAT for having considered TRAI's recommendations and passing the impugned order of August 30, 2007, contrary to statutory provisions.[18] It turned next to the authority of cases cited by the licensees, including *Cellular Operators Association of India* and *Delhi Science Forum*, and opined that they did not apply to the case before it. On the other hand, it cited a long line of cases to note that the Supreme Court had "consistently taken a view that once a licensee has accepted the terms and conditions of a license, he cannot question the validity of the terms and conditions of the license before the Court". The judgment held that TRAI and TDSAT did not have the jurisdiction to decide on the validity of the definition of AGR and to exclude items of revenue included therein in the license agreement between licensor and licensee.

On the third question of law on whether TDSAT's order had become binding on the government for not filing appeal against some of the licensees, the judgment held that the order of July 7, 2006, was not binding on the Union of India even in those cases where an appeal had not been filed.

On the fourth and final question of whether a licensee can challenge the computation of revenue and, if so, at what stage and on what grounds, the Supreme Court held that a licensee can raise a dispute if it believes a demand made on the basis of the computation of AGR is not in accordance with the license agreement after the license agreement had been entered into and when a particular demand is made by the licensor on the licensee. The judgment held that when such a dispute is raised against a particular demand, the tribunal would have to examine the relevant facts and the material basis for such demand and whether the demand "is in accordance with the license agreement and in particular the definition of Adjusted Gross Revenue in the license agreement and can also interpret the terms and conditions of the license agreement". On these arguments, the judgment allowed the government's appeals and set aside the impugned order dated August 30, 2007, of TDSAT. The court also set aside certain other orders of the tribunal that were based on the impugned order that had been set aside by the judgment and remitted them to TDSAT to pass fresh orders in accordance with law.[19]

The AGR judgment of the Supreme Court is significant for reining in the tendency of the specialized tribunal to assume the mantle of expert regulator for determining controversies of fact. The TDSAT judgment of July 7, 2006, was noteworthy on three counts: (1) for holding that it had the jurisdiction to decide disputes arising from licensees' challenge to the validity of the terms and conditions of the license, (2) for holding that effective consultation with TRAI entailed the obligation on the part of the central government to place before the regulator any material it intended to rely on in arriving at its final decision on TRAI's recommendations, and (3) for remanding the matter to TRAI for making recommendations directly to the tribunal and in the process inverting the roles of regulator and policy-maker by requiring the central government to present the material relied upon by it in rejecting its earlier recommendations. In arriving at these conclusions, the tribunal relied extensively on its interpretation of the Supreme Court judgments in the limited mobility case and in *Delhi Science Forum*. The limited mobility case, according to the tribunal, vested it with jurisdiction wide enough to decide on any dispute between licensor and licensee, and DoT's arguments that terms of contract could not be rewritten by it were met with the assertion that telecom sector licenses were different from 'ordinary' commercial contracts in that the right of the licensor to impose terms and conditions of contract was controlled by ITA and policy. On the crucial importance of policy, the tribunal drew upon *Delhi Science Forum*'s endorsement of NTP 1994. TDSAT also read the Supreme Court's judgment in *Cellular Operators Association of India* in conjunction with its own orders in *Petition no. 5 of 2002* to construe the terms of statute as imparting so much weight to TRAI's recommendations that their rejection was to be an exception and not the rule. Thereafter, in its judgment of August 30, 2007, the tribunal compounded its interpretation of the statute – which at least was dignified by being a judicial function – by actually deciding on the TRAI's recommendations on different items of revenue and costs that can be seen to be an egregious case of overreach unsupported by statute. These errors were made by the tribunal since it positioned itself as not merely an adjudicator of disputes but as the ultimate arbiter on matters of fact in relation to the expert regulator's recommendatory functions. This role confusion could perhaps be attributed proximately to a conflation of the provisions of section 14 (a) of the Act that vest adjudicatory powers on it in disputes between different sets of players and those of section 14 (b) that require the tribunal to hear and dispose of appeals against any direction, decision, or order of TRAI. The conflation itself resulted from a self-perception of TDSAT's role-criticality that was engendered by the Supreme Court's pronouncements in the limited mobility case. This conclusion is strengthened when it is noted that the two TDSAT orders were given under two different chairmen – Justice N S Hegde and Justice Arun Kumar – and in fact drew upon the history of assertive adjudication previously resorted to under Justice D P Wadhwa. The Supreme Court's judgment of October 11, 2011, applied a corrective on this tendency of TDSAT to go beyond the four corners of the statute and thus marks an important stage in the narrative of judicialization of regulatory governance in the sector. By restating the law related to the scope of TDSAT's jurisdiction in licensor and licensee disputes – especially arising from the licensor's disagreements with the regulator's recommendations given under section 11 (1) – the highest court once again not only altered the internal balance within the triadic dispute resolver but also clarified the ambit of the recommendatory power of the regulator and whether expert recommendations that were meant to aid the licensor in making policy level decisions were amenable to the tribunal's oversight, other than on terms allowed by a general judicial review.

3.4 Judicialization: summary and conclusion

Table 3.4 summarizes the chronology of key moments in the judicialization of the telecom sector in India as a prelude to the assessment of the role of each of the three branches of government in the initial constitution of the IRA and, subsequently, to evaluate the nature of relationships between them and " 'the fourth branch of government' " (Thatcher and Stone Sweet 2002: 16) that they collectively delivered as a mechanism to oversee the functioning of the telecom sector. As could

Table 3.4 Chronological summary of judicialization events

	Legislature	*Judiciary*	*Executive*
July 1991			New economic policy
January 1992			Metro CMTS tenders
February 26, 1993		Delhi High Court judgment in metro tenders case	
May 13, 1994			NTP 1994
July 26, 1994		Supreme Court judgment in *Tata Cellular*	
September 1994			Basic services guidelines promising TRAI
November 1994			Metro CMTS licenses issued
January 16, 1995			Circle CMTS and basic tenders
May 26, 1995			Indian Telegraph (Amendment) Bill prepared
June 2, 1995	Bill introduced in Lok Sabha		
August 9 and 16, 1995	Lok Sabha debate and passing of bill		
August 22, 1995	Amendment bill not taken up for consideration in Rajya Sabha		
November 23, 1995	Notice for introduction of TRAI Bill in Lok Sabha		
November 28, 1995	Motion for withdrawal of amendment bill in Rajya Sabha not pressed by government		
December 1995			Circle CMTS licenses issued
January 27, 1996			First TRAI Ordinance
February 19, 1996		*Delhi Science Forum* judgment of Supreme Court	

(*Continued*)

Table 3.4 (Continued)

	Legislature	*Judiciary*	*Executive*
February 27, 1996	Bill to replace ordinance introduced in Lok Sabha		
March 27, 1996			Second TRAI Ordinance
July 23, 1996	TRAI Bill, 1996, introduced in Lok Sabha		
November 8/21, 1996	Standing Committee report prepared/ presented to Lok Sabha		
January 19, 1997			DoT revises fixed-cellular intra-circle tariff
January 25, 1997			Third TRAI Ordinance
February 19, 1997			TRAI setup
March 3, 1997		Delhi High Court transfers disputes to TRAI	
March 15, 1997	TRAI Bill, 1997, introduced		
March 18/20, 1997	TRAI Bill passed		
March 28, 1997	TRAI Act		

	Legislature	Judiciary	Executive	Regulator
April 25, 1997				TRAI order in *Petition no. 1of 1997*
May 27, 1997				TRAI order in *Petition nos. 2 and 3 of 1997*
February 18, 1998				TRAI order in *Petition no. 3 of 1998*
June 1998			ICICI report	
July 16, 1998		Delhi High Court (Justice Usha Mehra) judgment in *UoI* v. *TRAI*		
November 9, 1998			BICP report	
November 20, 1998			GoT constituted	
March 9, 1999				Telecommunications Tariff Order (TTO) 1999
March 26/ April 1, 1999			NTP 1999	
April 1999			TRAI's recommendations on revenue sharing sought	
May 28, 1999				Telecommunication Interconnection (Charges and Revenue Sharing) Regulation (TIR) 1999
June 16, 1999			Attorney general's opinion obtained on industry distress	

(*Continued*)

Table 3.4 (Continued)

	Legislature	*Judiciary*	*Executive*	*Regulator*
July 6, 1999			Migration package approved	
July 22/August 1, 1999			Migration package offered/made effective	
July 1999			TRAI's recommendations on revenue sharing sought	
September 17/ November 1, 1999				calling party pays (CPP) regime announced with amendments to TTO/ TIR
October 12, 1999		Delhi High Court judgment in *Microwave Communication* affirming *UoI* v. *TRAI*		
November 15, 1999				global mobile personal communications service (GMPCS) revenue sharing recommendations
December 13, 1999			GoT-IT constituted	
January 17, 2000		Delhi High Court judgment in *MTNL* v. *TRAI* quashing CPP		
January 24, 2000			TRAI (Amendment) Ordinance issued	

March 9, 2000	TRAI (Amendment) Bill, 2000, introduced in Lok Sabha			
March 13/15, 2000	TRAI (Amendment) Bill passed by Lok Sabha/ Rajya Sabha			
March 25, 2000	TRAI (Amendment) Act			
April 17, 2000	Tenth Report of Standing Committee criticizes migration package			
June 23, 2000				TRAI recommendations on CMSPs' revenue sharing, including AGR definition
May–October 2000			TDSAT constituted	
August 31, 2000				TRAI recommendations on FSPs revenue sharing, including AGR definition
October 9, 2000			DoT seeks reconsideration of TRAI's recommendations and recommendations on limited mobility	
October 24, 2000				TRAI reconsidered recommendations, reiterating AGR definition for CMSPs

(Continued)

Table 3.4 (Continued)

	Legislature	*Judiciary*	*Executive*	*Regulator*
October 31, 2000				TRAI reconsidered recommendations on revenue sharing, reiterating AGR definition for FSPs/CMSPs/GMPCS
November 2000	Thirteenth report of the Standing Committee on IT		GoT-IT reconstituted for limited mobility issue	
January 2001				TDSAT starts hearing cases
January 8, 2001				TRAI recommendations on limited mobility
January 22, 2001				*Petition no. 1 of 2001* on limited mobility filed before TDSAT
January 25, 2001			DoT's FSP guidelines issued	
April 7, 2001			Limited mobility issue referred to GoT-IT	
April 26, 2001			GoT-IT recommends limited mobility	
August 29, 2001	Standing Committee report criticizes revenue sharing regime			

March 25, 2002		First TDSAT order in limited mobility litigation (Justice Sen)
December 17, 2002	Supreme Court decision in *Cellular Operators Association of India* remanding limited mobility matter to TDSAT	
August 8, 2003		Second TDSAT order on limited mobility litigation (Justice Wadhwa)
September 27, 2003		TDSAT judgment in *Petition no. 5 of 2002*
July 7, 2006		First TDSAT AGR order
September 13, 2006		TRAI AGR recommendations pursuant to TDSAT's first AGR order
January 19, 2007	Supreme Court order in AGR matter dismissing DoT's appeal	
August 30, 2007		TDSAT second AGR order
October 11, 2011	Supreme Court AGR judgment	

Source: Compiled by author.

be expected, the story of regulatory governance reflects the dynamics of change at the NIE L2-L3 cusp between institutional environment and governance on the one hand and at the more micro-political level of the individual institutional actors on the other. If at first the change in the nature of the game impelled recourse to a nuanced negotiation rationality strongly influenced by economic considerations arising from the inexorable march of market forces, this change in the rules of the game had institutional consequences for conflict resolution and resulted in the establishment of an independent regulator as the preferred triadic dispute resolution (TDR) mechanism. If TRAI was seeded in the executive's realization of the need for expert advice to mimic the marketplace, its birth as a creature of statute was midwifed by the judiciary, which recognized the need for a third-party dispute resolver to implement the changed rules of the game. The initial period from around the launch of the new economic policy in 1991 until the announcement of NTP 1994 and beyond was marked by intra-executive wrangling over the nature of regulatory governance in the sector and what role – if at all – should be assigned to the IRA vis-à-vis DoT. The judiciary provided gentle impetus to the recourse to expertise-based decision-making during this period as evidenced by the July, 1994, judgment in *Tata Cellular* with its preference for such expertise over a purely legal framework for governing contractual arrangements. The internal differences of opinion, however, continued to have a paralyzing impact on legislative efforts to craft a meaningful statute to provide the regulatory framework for the sector right until the February 1996 judgment in *Delhi Science Forum*.

Delhi Science Forum was important for forcefully coming down on the side of establishing an independent regulator. While granting the legislature's competence to set out sector policy both in terms of a vigilant Parliament's sovereignty over economic decision-making as well as the policy being compliant to pre-existing statute, the judgment also ranged over international experience in sector privatization to conclude that the creation of an independent regulator was critical for harnessing the private sector's appetite for private gains for the greater public good. By literally forcing the executive to bring in the TRAI Ordinance, 1996, this judgment marks the judiciary's acceptance of a particular form of economic rationality in the ordering of property and contractual rights and its insistence that this form of non-judicial legality be formally institutionalized to offer credible commitment to the private investor.

The exogenous application of judicial power had a mixed effect on the executive-legislature complex; while it created TRAI by ordinance and made efforts to pass the statute in Parliament, the period from 1996 to 1997 saw differences of opinion within the legislature over how the law ought to be framed. While some of the suggestions of the Parliamentary Standing Committee came to be incorporated, many others were not included in the final version of the bill that became the TRAI Act, 1997. The deliberative process of Parliament in this sense did not yield a robust act to govern sector regulation, as was soon evidenced in the way the TDR mechanism failed to produce the level of certainty required by the new private participants. TRAI was certainly guilty of regulatory overreach during the early years of its creation, especially on the adjudicatory side, but that excessive enthusiasm can be attributed to – if not exactly explained by – faulty design of the statute. The executive's lingering reluctance to share regulatory space with TRAI added to the crisis of confidence in the framework's ability to offer a credible commitment to achieve the stated policy goals. The three years of TRAI-Mark I to 2000 saw the regulatory mandate being severely tested, and the role of the higher judiciary – in the form of the Delhi High Court judgments in *Union of India* v. *TRAI* (July 1998) followed by its quashing of the CPP regime in January 2000 – was again significant in pushing the legislature to bring in the amendment in 2000. Categorizing the regulator's functions broadly into adjudicatory, recommendatory, and mandatory – with individual variations across different judgments – was a legacy of the judicial pronouncements of this period, and these categories were carried into not only later courtroom language, but also the way the executive (in the amendment ordinance) and the legislature (in the Act that followed) construed and articulated the different functions of the sector regulator. The separation of the adjudicatory functions into TDSAT following the amendments made to the entire regulatory governance regime is particularly significant since with the invention of the specialized tribunal as part of the sector's governance structure (rather than as part of the higher judicial edifice) the judicial mentality came to be firmly embedded in the IRA as a corrective to excessive reliance on technical expertise. This layering of judicialization on to the TDR mechanism represented the judicial urge to apply a corrective to excessive reliance on narrow technical expertise. At the same time, in retaining TRAI as essentially a recommendatory body, the amending Act did not go far enough in institutionalizing independent regulation as an arena distinct from the executive's policymaking process (Tangirala 2015: 59).

It is interesting that TDSAT initially could not shed the judicial reluctance to interfere in expertise-based decision-making. Its deference to executive discretion was starkly brought out in its judgment in *Petition no. 1 of 2002* in March 2002 on the question of limited mobility in which it kowtowed to the executive's policy direction, which was also endorsed by TRAI albeit under duress. It once again fell to the Supreme Court – in the *Cellular Operators Association of India* judgment of December 17, 2002 – to clarify the balance of judicial and discretionary powers between TRAI and TDSAT and to reinvigorate the judicial mandate of TDSAT. The apex court's aggressive widening of the powers of the specialized tribunal that would act as its agent in the sectoral IRA had a salutary effect on the way litigation was conducted in and the disputes resolved by TDSAT during the crucial years marked by a considerable number of section 14 (a) disputes that followed. By virtue of this judgment, the roles of TRAI, TDSAT, and the Supreme Court with respect to dispute resolution in the telecom sector were placed in perspective. The remarkable independence showed by TDSAT thereafter even in cases where majority judgments favoured the executive line of thought is proof of the positive effects of judicialization.

The fine balance sought by the judgment, however, was disturbed over time by the TDSAT's overreach of its adjudicatory functions. This tendency – visible even as early as in the judgment of September 27, 2003, in *Petition no. 5 of 2002* – was exemplified by its judgments in the AGR case on July 7, 2006, and August 30, 2007. In these judgments, TDSAT was guilty of not only aggrandizing to itself the discretionary powers that ought to be legitimately left to TRAI but also of transgressing the bounds of statute in directing the licensor to present its views before the regulator in a dispute raised by licensees, thereby inverting the entire regulatory process flow. Added to this was TDSAT's legal misreading of the ambit of licensor-licensee disputes. It was this excessive enthusiasm, along with the conflation of section 14 (a) and section 14 (b) functions, which was sought to be curbed by the Supreme Court's AGR judgment of October 11, 2011. By reminding the specialized tribunal of its limitations – and doing so by recourse to the rather bland but workmanlike formulation of questions of law as mandated by section 18 of the TRAI Act as well as its own earlier decisions – the Supreme Court restored the balance that had swung too far to the side of the tribunal, which saw for itself a role in directing the licensor's *and* the regulator's exercise of discretionary power in one particular direction. While at first glance the Supreme Court's judgment may seem to disfavour the licensees' interests, by locating the effect of the judgment in the longish judicialization narrative it became obvious that, on the contrary, the judgment can be expected to facilitate the rule of law in the telecom sector in the long run. Ironically enough, the Supreme Court's October 2011 judgment effected this rebalancing by reducing the shadow cast by its own judicializing agent over the exercise of non-judicial legality by TRAI.

Notes

1 This aspect was also noted by the Standing Committee in its report when it stated that "the only immediate effect of the Ordinance was the termination of the terms of office of the Chairman and Members of the erstwhile TRAI" (SCIT 99–00 2000: 30).

2 Justice Sen was also president of the National Consumer Disputes Redressal Commission until February 29, 2001. See websites of TDSAT (www.tdsat.nic.in) and National Consumer Disputes Redressal Commission (NCDRC; see ncdrc.nic.in).

3 Later, the power of the Supreme Court under section 18 of the act was held to be greater than its power of judicial review when it was observed that since in appeals under section 18, "this Court will be concerned with a substantial question of law arising in the case, its jurisdiction would not be restricted to illegality, irrationality or procedural impropriety in the decision making process" (*Cellular Operators Association of India and Ors.* v. *Union of India and Ors.* (2003) 3 SCC 186).

4 Section 11 (4) of the TRAI Act, stipulating that TRAI "shall ensure transparency while exercising its powers and discharging its functions", is a provision contained in the pre-amendment act as well, and it entailed a wide consultation process before making recommendations, which was probably expected to take considerable time.

5 The reference to the attorney general was obtained, strictly speaking, as a measure to follow up on the cabinet decision to make NTP 1999 uniformly applicable throughout the country. This was the justification DoT offered to various constitutional bodies (the president, the Election Commission) when the propriety of the decision was subjected to debate (see JPC 2013: 22, 26).

6 The terms were communicated to cellular licensees in DoT's letter no. 842–153/99-VAS (Vol. V) (Pt.) dated July 22, 1999. A similar package was offered to the six basic service licensees as well, extending their license tenure from 15 to 20 years.

7 The gross revenue was to be calculated by deducting PSTN related call charges paid to DoT/ MTNL and service tax collected on behalf of the government by the licensee from subscribers.

8 The CAG report argued that license tenure for cellular licenses was already extended from 10 to 15 years based on the ICICI report and that TRAI had increased the rentals from Rs 156 to Rs 600 per month with effect from May 1, 1999, after the BICP report containing the same recommendation was forwarded to it by DoT in November 1998 (CAG 2000: paragraph 14.7.2). DoT announced the extension of cellular license period from 10 to 15 years on October 8, 1998 (Uppal et al. 2006: 21).

9 One licensee in Uttar Pradesh (West) circle gave the unconditional acceptance only in November 1999 and did not pay the dues. Its license was terminated.

10 The DoT sought TRAI's recommendations on revenue sharing arrangements in tranches and in different contexts. For example, the request for recommendations on license fee arrangements for migration of cellular licensees to the new regime was sought on July 12, 1999, which followed earlier requests on April 23, 1999, and October 7, 1998, for appropriate level of license fee, revenue share percentage, definition of revenue, etc. under NTP 1999 and quantum and structure of license fee payable by circle cellular licenses for the extended period of license from the 11th to the 15th years respectively (see TRAI 2000a). In the case of basic service licenses, too, two references were made by DoT in April and July 1999, the first seeking license fee recommendations on the issue of fresh licenses for FSPs in 15 circles and the second for similar recommendations for the migration of six basic service licensees to the new regime (see TRAI 2000c). Both were preceded by a recommendation made on November 15, 1999, by TRAI in response to a DoT request for arriving at licensing terms for a global mobile personal communications by satellite (GMPCS) license, which can be considered TRAI's first attempt at defining revenue sharing arrangements (see TRAI 1999e).

11 The tariffs for cellular mobile services were still quite high at this stage.

12 This conflict had two sources at this point: one, the change in business dynamics because of allowing limited mobility to fixed subscribers when the cellular industry was still in a nascent stage of development, and two, the government's stand articulated a year previously that "it considered such mobile services using a handset unacceptable, since they would overlap with services of licensed cellular operators" (Uppal et al. 2006: 8). It has been noted that DoT and CMSPs had "a similar interpretation" of the WLL provision in NTP 1999 until August 2000 (Mukherji 2009: 508). Limited mobility became the thin edge that later led to further policy changes and ultimately to unified access service (UAS) licenses.

13 Before the TRAI recommendations of January 8, 2001, were to be considered by the Telecom Commission on January 24, 2001, the Cellular Operators Association of India sought a direction from TDSAT in a petition filed on January 22, 2001, that "the Telecom Commission should not consider the recommendations of TRAI on the issue of limited mobility but no such direction was issued by [TDSAT]" (GoT-IT 2001: 4).

14 The fixed line network was divided into SDCAs and LDCAs (long distance charging areas) under the Indian Telegraph Rules, 1951. SDCA was coterminous with the local area for charging for trunk calls. Each SDCA covered anywhere between 800 and 2,000 square kilometres and followed administrative divisions. Every SDCA was allotted a unique trunk code, and calls within it required the dialing of only the subscriber number without prefixing the long distance access code ('0') (see National Numbering Plan [NNP] 2003).

15 The January 2001guidelines (no. 10–2/2000-BS-II) are notable for many reasons, most importantly for the legacy they established for bundling and charging spectrum. In this section, the discussion is restricted to the permission granted for limited mobility.

16 The constitution of GoT-IT was brought to the notice of TDSAT by the petitioners in *Petition no. 1 of 2001*. After hearing parties, it ordered on

April 17, 2001, that it would await the decision of the government before passing any order (GoT-IT 2001: 31).

17 The first two cellular licensees in telecom circles were allowed a reduction of another 2 per cent for four years starting from April 1, 2004, subject to a floor of 5 per cent that constituted the universal service obligation levy, in another instance of considerable interest for students of political economy.

18 The judgment also noted that the central government in any case was not agreeable to the fresh recommendations of TRAI, and its decision not to alter the definition of AGR was final under the first and fifth provisos to section 11 (1).

19 In a supplementary order on the same date, the Supreme Court also gave licensees two months' time to raise disputes in accordance with its holding on the fourth question of law and said that during these two months, the demands already raised by DoT would not be enforced. As could be expected, most of the demands were then contested by the licensees before TDSAT during this period. The cases have again reached the Supreme Court in a second round of litigation.

Part III

CRAFTING ORDER?

Two decades of the regulatory institution

This section traces the history of the Telecom Regulatory Authority of India (TRAI)'s regulatory journey over the years and assesses its performance in achieving the objective of crafting order in the regulatory space. It divides and conducts the evaluation in three phases: the first phase prior to the amendment of the TRAI Act, the second between the years 2004 and 2007 by which time unified access became a reality, and the final third phase beginning from 2008. The discussion encompasses licensing aspects, limited mobility, tariff setting, interconnection arrangements, transition to unified access, entry fees, spectrum allocation and pricing, the shift to hypercompetition, capping of access service licenses, and the litigation that followed, which became known as the 2G case. It discusses the implications of the judgment in the case and the opinions in the subsequent presidential reference on the spectrum allocation regime in India.

4

THE FIRST REGULATORY PHASE

4.1 Typology

There is no gainsaying the influence exerted by the courts on the trajectory of regulatory evolution in the Indian telecom sector. The early developments in the sector carry the unmistakable imprint of judicialization, which mediated the inevitable economic contestation over policy choices to put in place a triadic dispute resolution (TDR) mechanism that balanced the need for stable property rights and the imperatives of government oversight. The description of the construction of a two-part independent regulatory authority (IRA) that has occupied our attention so far is analytically important to gain perspective on the mode of dispute resolution in the sector and the way in which its evolution was shaped by the 'judicial' ordering of dyadic conflicts. It allows us to discern the limits of expertise-based decision-making at the gates of a more generalized legal arena where regulatory decisions are tested against constitutional principles related to property rights and contracts but which is nevertheless situated within the broader sectoral locale to account for its distinctive techno-economic features and peculiarities. At the same time, it is necessary to examine the impact of judicial intervention on the Telecom Regulatory Authority of India (TRAI) and whether the role of the courts has helped or hindered the non-adjudicatory functions of the IRA that were vested with the expertise-based regulatory decision-maker. Assessment of TRAI's performance as independent regulator needs to move beyond a merely descriptive account of its activities over the years to an examination of its institutional evolution from its initial establishment in 1997 as a potentially omnipotent entity albeit with an undifferentiated functional mandate to its assertion of a primarily recommendatory role under the oversight of the Telecom Disputes Settlement and Appellate Tribunal (TDSAT) and the higher judiciary subsequent to the amendment to the statute in 2000 and onwards to the ignominy of the observation from the Supreme Court in 2012 that "the recommendations made by TRAI were flawed in many respects". The process of vesting broader discretion in the independent regulator by the legislature-executive combine was accompanied by an insistence of adherence to fair process, procedural safeguards, and consistency by the judiciary in the regulator's decision-making functions. As such, what has been seen as a "working accommodation" between administrators and judges in the true spirit of the traditional model of administrative law (Stewart 1975: 1681) is also amenable to being described as judicialized regulatory practice.

TRAI's non-adjudicatory role may usefully be described in terms of three phases, even though no clear disjunctions mark the transitions between periods. The two primary axes of differentiation are competition and growth in number of telephone connections since 1997. It is possible along these two dimensions to posit the first phase of regulatory activity of TRAI from 1997 to 2003 (preparatory), the second phase from 2004 to 2007 (competitive), and the third, ongoing phase from 2008 (hypercompetitive).[1]

4.2 First phase of regulation: 1997–2003

The first phase of telecom regulation in India was driven by the imperatives of competitive service provisioning. In terms of the TRAI Act, this preparatory phase may be divided into the pre-amendment and post-amendment periods, with substantially different boundaries set on the regulator's functioning in each. The way TRAI conducted adjudication in the pre-amendment stage – especially in the first year of its functioning that led to the momentous decisions by the Delhi High Court which precipitated policy change – reflected the problematic nature of the delegation of regulatory responsibility to the regulator under the statute. The absence of classification of the varied functions assigned to the regulator and the haphazard lumping together of disparate functions pertaining to dispute resolution, sector regulation, and making recommendations to the policymaker were rectified by the courts, which sought to categorize TRAI's functions along these different dimensions. The courts' division of TRAI's functions into adjudicatory, recommendatory, and mandatory categories had considerable influence on the legislative branch as well, and the amended Act incorporated these categories while farming out the adjudicatory functions to TDSAT and distributing TRAI's recommendatory and mandatory functions among sub-sections (a) and (b) of section 11

(1). TRAI nevertheless functioned for more than three years under the pre-amended act, and the absence of statutory clarity demonstrably affected the way it discharged its non-adjudicatory functions during this period. The following analysis attempts to show, in turn, that the legacy of pre-amendment regulatory decision-making continued to affect the conduct of regulatory affairs well into the second period of the first phase and even in the later phases.

Regulation in the pre-amendment period

TRAI's agenda in its first year of existence was dominated by telecom pricing issues for basic services, even though other aspects such as universal service obligation, interconnection, numbering plan, internet, registration and performance monitoring of service providers, tariff notification applications by the Department of Telecommunications (DoT), some paging service operators and cellular service providers, and interaction with consumer organizations too were addressed, apart from performing its adjudicatory functions (TRAI 1999a). This was no doubt influenced by the unambiguous mandate of section 11 (2) of the Act. TRAI also considered the determination of equitable tariffs as singularly important for the reason that "significant private investments can only take place if adequate return is possible for investments in the various parts of the network" (ibid: 11),[2] the various parts being (1) local, (2), domestic long distance, and (3) international telecom services. It bears notice that competitive service provisioning existed only in the local services market when TRAI started functioning; India's World Trade Organization (WTO) commitments entailed review of domestic or national long distance (NLD) and international long distance (ILD) telecom services in 1999 and 2004 respectively. DoT was the monopoly domestic long distance service provider, and the public sector undertaking Videsh Sanchar Nigam Limited (VSNL) was the monopoly provider of international long distance services at this time. It is also necessary to note that the incumbent's landline network was about 18 million strong in March 1998, while the cellular service providers' subscriber base was still only about 0.85 million[3] (ibid: 27).

Very early in its existence, TRAI released its first consultation paper on telecom pricing in November 1997 and discussed concepts, principles, and methodologies as well as options for telecom tariff reform. This was followed by a second consultation paper released in September 1998 "which proposed a framework for telecom tariff policy and suggested specific tariff options for various Telecom Services"

(TRAI 2000b: 29). TRAI's tariff rebalancing efforts had two prongs: (1) aligning retail prices of various parts of the network to costs in two steps and (2) aligning interconnection charges to costs. Both aspects were critical to implementing its key economic function of ensuring a measure of credible commitment to potential private sector investors in the Indian telecom sector. Retail prices of services and interconnection charges applicable inter se to service providers were also closely intertwined and were additionally linked to how revenue sharing arrangements among service providers for calls borne on multiple networks would eventually be determined.

The telecom pricing exercise conducted by TRAI led to the issuance of the Telecom Tariff Order (TTO 1999) on March 9, 1999, and the Telecom Interconnection Regulation (TIR 1999) on May 28, 1999. Together, they represented the economic philosophy animating the regulator's approach and radically altered the way costs were translated into consumer prices. At the same time, TRAI's approach itself was influenced by the nature of disputes that it encountered between the established incumbent (DoT) and the fledgling private sector access service providers. Chapter 2 noted that the nature of disputes brought before the regulator around this time centred on the investment distress experienced by licensees labouring under an unequal contractual relationship and non-vestment of clear-cut dispute resolution functions in the IRA. DoT, in its policy-making role and supported by the Parliamentary Standing Committee, resisted TRAI's attempts to reduce cross-subsidies in the local network derived from inflated long distance prices that also allowed it to garner a substantial surplus "on the grounds that [the surplus] was needed for expansion of the telecom network" (TTO 1999: 53).[4] At the same time, in its service provider role, DoT exploited its essential facilities to block organic expansion of the new entrants' networks by allowing limited interconnection as well as by charging whimsical tariffs that had little nexus with costs. To complicate matters, the new entrants from the private sector – both landline and cellular licensees – also faced problems of financial closure in view of their unrealistically high license fee bids that starkly overestimated market potential. Their unequal position vis-à-vis DoT, while certainly contradictory to the government's stated aim of inviting private sector participation in telecom service provision, also offered the licensees an opportunity to renege on essential reciprocal commitments written into the contract.

It is necessary to sketch in some detail the outcomes of the bidding processes that led to the licensing of basic and cellular service licenses in the first round of competitive service provisioning to better appreciate the framework under which TRAI began its regulatory practice. Table 4.1 gives the salient events in this round when the first cellular mobile telephone service (CMTS) and basic service licenses were issued to private sector players.

Other than the unorthodox sequencing of regulatory reform, what is striking about the opening up of the Indian telecom services sector was the licensing of access service licenses first without concomitant de-monopolization of long distance services. The selective introduction of competition in the access network, which was cross-subsidized by the incumbent using surpluses from long distance network revenues, led directly to the incorporation of tariff rebalancing at the top of the regulator's agenda as a surefire means of signalling credible commitment. The imperative for tariff rebalancing was further strengthened by national commitments at WTO to review domestic and international

Table 4.1 First round of licensing

Date	*Event*	*Remarks*
January, 1992	Technical bids invited for metro CMTS	Duopoly*; beauty parade**; fixed license fee; ceiling on tariffs
November, 1994	10-year metro CMTS licenses (Group Speciale Mobile, GSM) issued	Separate spectrum usage charges for maximum 4.5 MHz in 900 MHz band
January, 1995	Bids invited for circle CMTS and basic service licenses	Duopoly*; single stage tender (auction)***
December, 1995	10-year circle CMTS licenses (GSM) issued	34 licenses in 18 circles; no bids/one bid in two circles each; maximum 4.4 MHz in 900 MHz band
September, 1997/ March, 1998	15-year circle basic licenses made effective	One license each in six circles

* Duopoly in each market entailed two private operators in CMTS (with DoT reserving the right to enter as third CMTS operator) and one operator in addition to DoT in basic service.

** The selection was based on the following criteria: company track record, technology compatibility, utility of technology for future development, protection of national security interests, affordable quality of service (QoS), and attractiveness of commercial terms offered, later incorporated in NTP 1994 as well.

*** The tender for CMTS was given to the highest bidder (subject to technical qualification), and the second highest bidder was to match the highest bid; for basic services, the tender evaluation included weights for bid amount (72 per cent), network rollout (10 per cent), rural connections (15 per cent), and indigenous equipment use (3 per cent).

Source: Compiled by author.

long distance service liberalization by 1999 and 2004 respectively. It could even be speculated that the impending review of competitive service provision in the long distance markets was one of the factors that led to enthusiastic participation from international majors (often in partnership with large Indian conglomerates)[5] and the unrealistically high bids on the part of license seekers in the access service tenders that allowed them a foot in the door of the potentially lucrative 'full service' telecom market. Yet another effect of this policy level decision of the government was the institution of inherent conflict in the dyadic relationship between DoT – in its role as a full service provider – and the new entrants whose licenses permitted them to provide only access services and to compulsorily seek interconnectivity from the incumbent's long distance network. This structural imbalance in the licensing process intensified the already significant incumbency advantage DoT enjoyed because of network externalities and forced an adversarial relationship between it and the private sector participants – as, for example, in determining equitable access charges for use of DoT's inter-circle and international long distance networks.[6] This had a direct impact on the regulatory performance of TRAI, especially in the first phase of sector regulation and in the manner it went about its tariff-setting function.

A second feature of interest in the process of privatization was the introduction of competition in the cellular market before opening up the basic service market. Notwithstanding the litigation that led to the *Tata Cellular* judgment, that the licensing of cellular services in the metro circles could be carried out at all can be attributed to the absence of significant political opposition to a service that was still in its infancy and, more crucially, one that was not offered by DoT itself. The licensing of cellular services – lumped under the omnibus term 'value-added services' along with electronic mail, voice mail, data services, audio text services, video text services, video conferencing, and radio paging – represented a tactical lapse on the part of the otherwise intransigent bureaucracy in DoT, which obviously did not foresee the disruptive potential of the mobile phone on the telecom services ecosystem at that point in time. The privatization of cellular mobile services in the metros – which began in January 1992 and led to the issuance of licenses in September 1994 – proved to be the thin edge of the wedge for the later expansion of competitive access service provisioning in the Indian telecom sector when the tenders for circle cellular and basic service licenses were announced in November 1994 following NTP 1994. It transpired, of course, that the tenders for circle cellular mobile licenses were not only concluded much faster (by December 1995) but were also far more successful than those for basic services (that were effective from September 1997/March 1998), where only 6 licenses were issued compared to the 34 issued for cellular operators.[7] The markets discerned what DoT did not foresee, although the process of litigation in *Delhi Science Forum* certainly helped delay the investment decisions. The first mover advantage accrued to cellular mobile service providers, coupled with the rapid technological advances that brought down capital costs of cellular networks, was jealously defended by the CMSP lobby (the Cellular Operators Association of India, COAI), which came under attack from the later basic service licensees who formed the Association of Basic Telecom Operators (ABTO) and took the battle for increased access to the access services market by using wireless in local loop (WLL) to offer limited mobility services before regulatory, executive, and legislative forums with considerable enthusiasm.[8] This trajectory of opening up the access services market – whereby cellular licenses had a clear head-start but were also in competition with some private sector basic service providers who commanded considerable financial and managerial resources even as they began telecom service provisioning later – had a significant later impact on how TRAI responded to its regulatory mandate when it gave recommendations first on limited mobility in basic services in January 2001 and subsequently on unified access service (UAS) licenses in October 2003 under the changed policy regime after NTP 1999. TRAI's recommendations in these matters – with modifications – formed the basis for the licensor's decisions that significantly affected the competitive scenario in the access service market that in turn fed into the dispute cycle before TRAI.

A third salient feature of the first round of privatization was the grant of licenses to the highest bidder through single-stage auction. Even though the Supreme Court upheld the government's approach for the grant of metro CMTS licenses, which led directly to the incorporation into policy of selection criteria for tender evaluation in NTP 1994, the 'beauty parade' system with pre-determined license fee used for awarding licenses in the metros was dispensed with in the tenders for circle licenses, which were instead awarded on the basis of highest bid license fee. This auction, while "spectacularly successful" (Sridhar 2012: 78) in terms of bids received, in fact represented the new entrants' anxiety to enter the potentially lucrative Indian telecom market. The subsequent market realities –including a skewed regulatory institutional framework – forced the bidders to renege on their commitments, leading directly to a change in policy and to the migration package that replaced the fixed license fee regime with a revenue sharing arrangement. It is also interesting that while the clamour for mitigation of license fee was largely based on the distress experienced by CMTS providers, the basic service operators were also recipients of the ameliorative migration package even before their networks became fully operational and their profitability could be objectively assessed. This 'follow on' effect had a role to play in worsening the turf battles between COAI and ABTO, with consequential effects on regulatory governance of the sector. Curiously, government as policy-maker ignored TRAI in the process of arriving at the decision regarding the migration package even as it based its decision on a number of reports from other entities such as Bureau of Industrial Costs and Prices (BICP) and Industrial Credit and Investment Corporation of India Limited (ICICI), an empowered body (GoT), and a constitutional authority (the attorney general for India) who gave inputs supporting the change of policy that led to the migration package and NTP 1999.[9] This significant omission of consultation with the sector regulator has not been noticed in the literature that essentially viewed the changeover as an instance of policy change.[10] However, given the section 11 mandate of TRAI even under the pre-amended Act to recommend need and timing for introduction of new service providers (section 11 (1) (a)), recommend the terms and conditions of license to service providers (section 11 (1) (b)), ensure compliance of the terms and conditions of license (section 11 (1) (e)], recommend revocation for non-compliance (section 11 (1) (f)), facilitate competition and promote efficiency in the operation of telecom services to facilitate growth (section 11 (1) (h)), and protect consumer interest (section 11 (1) (i)), the complete absence of consultation with TRAI on the migration package, involving directly or indirectly all these sub-sections of the statute, is as good an indicator as any of the stature it commanded within executive circles in the first years of its existence. While the Delhi High Court judgment in *Union of India* v. *TRAI* 1998 Indlaw Del 562 of July 1998, which held TRAI's recommendatory functions under sections 11 (1) (a), (b), and (f) as not 'regulatory' and therefore as not binding upon the licensor, who is not obliged by statute to seek such recommendations before exercising its plenary, quasi-judicial powers under section 4 of the Indian Telegraph Act (ITA), is often blamed for the emaciation of TRAI, the judgment did not preclude the government from consulting TRAI before finalizing the package to be offered to telecom access services licensees who were subject to the regulator's jurisdiction in any case. That the government – in this case extending beyond the usual suspect, DoT, which had in any case adopted an adversarial stance vis-à-vis TRAI – did not choose to do so reflects the lack of institutionalization of independent regulation during the early years of the first phase of regulatory governance. That this non-consultation did not evoke strong comment or criticism hints at a larger consensus among state actors and independent observers regarding the efficacy of referring complex regulatory matters to TRAI.

Thus – once again – TRAI had no role to play in setting the rules of the game that were altered significantly between NTP 1994 and NTP, 1999. It had to be content with the professed policy of government to arrive at arrangements to govern the modified contracts after consultation with TRAI, one of the objectives of the subsequent amendment that also reconstituted TRAI in January 2000. It was a chastised and demoralized TRAI that accepted without demur the NTP 1999 stipulation that government would consult TRAI for recommendations on the "appropriate level of entry fee and percentage of revenue share arrangement". That TRAI was not involved in the much larger institutional issue of resetting the rules of the game and governance arrangements to ensure the orderly growth of the telecom sector but only in third-order economizing decisions related to revenue shares and entry fees (instruments to align incentives and encourage efficient resource allocation) is a marker of the subordinate status given to the IRA in the governance scheme. This secondary role had considerable implications for the way TRAI conducted regulatory affairs ever since and was reinforced by changes to the statute that relocated adjudicatory functions in a separate specialized tribunal and thereby reduced the formal stature of the organization in the executive hierarchy. Thus, the fact that TRAI focused its energies on tariff rebalancing in the early years is evidence not only of the recognition by the regulator that tariff reform was "key to the successful implementation of the government's policy objectives" (TRAI 2000b: 29), but also of the straitened circumstances of its institutional existence outside the core of executive decision-making. While aligning tariffs with costs was no doubt a powerful signal to investors that TRAI was capable of bringing expert economic input into regulatory decision-making, the exercise undertaken by TRAI for resetting tariffs also reflected a pragmatic acceptance of its limited mandate in practice, notwithstanding the letter of the statute. After all, it was only section 11 (2) of the TRAI Act that was never at issue in legal and statutory terms.

It must be noted that TRAI's tariff-setting efforts were not free of controversy. To return to the tariff-setting story, since the access services market was dominated by fixed phones, the consultation papers released by TRAI in November,1997 and September 1998 concentrated on proposals for rebalancing tariffs for basic services that would perforce lead to increases in monthly rentals and local call charges and a corresponding reduction in long distance call charges. This tariff rebalancing orientation faced opposition mainly on three grounds: (1) that there was as yet no competition in the basic services segment, (2) that DoT's surpluses could be used to lower tariffs for all services, and (3) that increase in rental charges would depress demand and consequently network expansion and growth (TRAI 2000b: 30). The resistance to change was widespread and vocal and even prompted the Parliamentary Standing Committee on Communications to examine the matter and make a report, which was presented to the Lok Sabha on December 2, 1998. One of its recommendations referred to studies that estimated that bi-monthly bills of subscribers who used their telephones to make mostly local calls were likely to rise by 52–72 per cent and 140–310 per cent in metropolitan cities and rural areas whereas business subscribers and affluent sections of society who made long distance calls more extensively would see reductions ranging between 30 per cent and 50 per cent; the report advocated that "as a matter of last resort, the Government may consider taking recourse to the provisions of Section 25 (2) of the Act" (Standing Committee on Communications [SCC] 99–00 2000a: 7). TRAI responded to these concerns by referring at length to – as well as trying to meet squarely – the arguments made in the report of the Standing Committee in its tariff order of March 9, 1999, that was to come into effect from April 1, 1999. While TRAI believed it had balanced "the various concerns, including mutually conflicting ones" (TRAI 1999b: 53), the government obviously did not, given that "soon after the announcement of the Tariff Order by TRAI, the steep increase in Rentals and local call charges invited sharp criticism from Public and Members of both Houses of Parliament" (SCC 99–00 2000a: 7), and a directive was issued to keep TTO 1999 in abeyance on the very day after it was issued.

The policy direction issued by the government under section 25 of the Act has been blamed on DoT, which was noted to have responded to TTO 1999 with "outrage and fury" (Raghavan 2007: 615). This may not be entirely accurate, given the reports of stiff opposition by several political parties, including a major member of the governing coalition that threatened to boycott Parliament unless the hike was withdrawn.[11] TRAI did not stand down, however, and once again joined battle with DoT, which responded in a letter dated March 27, 1999, with an ostensibly pacifist request to postpone the date of implementation. Some of the basic and cellular licensees also sought time for implementation of revised tariffs. After holding consultations[12] with service provider associations and DoT, TRAI notified the revised date of implementation of tariffs notified in three key schedules of TTO 1999 to be May 1, 1999 (TRAI 2000b: 31).

It was in this fashion that the pre-amendment period in the first stage of TRAI's existence came to be dominated by tariff related issues on the non-adjudicatory side. While tariff setting is an essential ingredient of regulatory governance in the early years of competition, rebalancing of basic services tariffs received exaggerated attention both from the regulator and within the larger political economy quite out of proportion with its importance at that point in time, when fixed telephony was on the threshold of losing its primacy to the cellular mobile telephone service. It is true that rebalancing was undertaken prior to opening up long distance services to competition as well, and TRAI's aggressive regulatory stance vis-à-vis DoT – itself a consequence of the policy decision to open up only access services to start with – emanated from the justified objective of controlling the horizontally integrated incumbent from exploiting its dominant position to the detriment of competition. The setbacks that TRAI faced on the adjudicatory side during this period fed into its regulatory zeal, albeit over a constricted span. Faced with repeated attrition of its regulatory mandate as enunciated in section 11 (1) of the pre-amended Act due to the courts' interpretation of its provisions, TRAI responded with the one instrument that was unambiguously available to it under section 11 (2) and fashioned it to undercut DoT's moves as service provider that denied fast, ubiquitous, and affordable interconnection to its competitors. What TRAI lost in litigation, it sought to make up in tariff rebalancing. The "regulatory mess" that had resulted from the court rulings between 1997 and 1999 (Mukherji 2004: 287) created a situation where TRAI performed its one clear regulatory function with single-minded devotion and brought large doses of economic expertise to what was hitherto considered a preserve of narrow technical proficiency. It did this with considerable success, and it can be observed that "the key elements of the framework introduced by the original Telecom Tariff Order in March 1999 have largely survived all the amendments" (Raghavan 2007: 615). Similarly, the principles that animated TIR 1999, which was the governance mechanism for payments for interconnection arrangements among service providers, were robust enough to endure the many challenges brought by the more complex networks of later years.

The close attention TRAI paid to tariff setting had unintended consequences as well. The successful execution of rebalancing and its positive impact has received much favourable attention, while its baleful effect on an 'unreformed' DoT has not. TRAI's primary objective was to prepare the ground for competition, and it used its tariff-setting powers to (1) in the case of fixed phones, reduce prices for long distance calls that were way above cost and to increase rental and local call charges that were significantly below cost, and (2) in the case of cellular phones, to restructure the prevailing low rental and high call charges "that dissuaded usage and loaded the subscriber base" (Kathuria 2007: 204). The other tariff initiative was to ensure that DoT's predatory policies – such as steeply increasing fixed-to-mobile call charges or denying adequate number of points of interconnection and thereby driving up cellular service prices – were contained. This twin-pronged approach severely affected DoT's viability as a service provider. Since privatization – or even corporatization – of the incumbent did not precede entry of competition in the telecom sector in the unorthodox Indian reform sequence, DoT had to simultaneously face the attrition in incumbency advantage due to tariff reform as well as all the usual rigidities associated with its position as a centrally ordered, hierarchical departmental service provider. This sowed the seeds for the decline and loss in profitability of DoT as service provider, a legacy-in-the-making it bequeathed to its public sector successor, Bharat Sanchar Nigam Limited (BSNL), which was formed on October 1, 2000. The opening up of the long distance sector to competition further affected BSNL's ability to garner surpluses. Because tariff reform remained the only item on the agenda that TRAI could meaningfully pursue, because TRAI pursued tariff reform doggedly and with considerable expertise, because DoT's survival instincts as a service provider lacked a long range perspective and were limited to stymieing TRAI's pursuit of legitimate regulatory activity, and because the government (of which DoT was an integral part) for valid political reasons did not undertake even the token first step of corporatizing the incumbent to prepare it for competition, the rebalancing exercise had long-term deleterious effects on the incumbent service provider. Shorn of its surpluses at an early stage in the competitive cycle and unable to restructure its operations to deal with nimbler adversaries in the marketplace, DoT as service provider was condemned to sub-optimal performance long into the future.[13]

Regulation between 2000 and 2003

The new regulatory framework post-amendment was tested when the government sought the regulator's views on the scope of the WLL technology that was permitted to basic service providers under NTP 1999, specifically on whether fixed service providers (FSPs) could provide mobility to consumers using the electromagnetic spectrum deployed in the local loop. Strictly speaking, the controversy over permitting 'limited mobility' of fixed line subscribers occurred at the cusp of policy change and statute amendment: NTP 1999 laid the groundwork to make the fixed service provider licenses more attractive. Among other palliative initiatives of NTP 1999 were those that stated that fixed service providers "shall be freely permitted to establish 'last mile' linkages to provide fixed services and carry long distance traffic within their service area without seeking an additional licence"; that interconnectivity between service providers in different license areas would be reviewed as part of the liberalization envisaged in long distance services in consultation with TRAI; that FSPs would be allowed to provide "all types of fixed services including voice and non-voice messages and data services, utilizing any type of network equipment" that meet specified standards; that separate 20-year licenses would be issued for each service area without limiting the number of licenses that an FSP could hold; that they would be required to pay a one-time entry fee and license fee in the form of revenue share to be determined in consultation with TRAI; and that the spectrum required by FSPs for WLL services would be made available on "payment of an additional one time fee over and above the FSP entry fee" and that they would pay license fee as revenue share for spectrum utilization, to be recommended by TRAI. Taken together, these policy stipulations in NTP 1999 represented a substantial incentive to floundering FSPs whose business plans were in danger of being disrupted by rapid advances in cellular mobile telephony. NTP 1999 attempted to balance these concessions given to FSPs by permitting cellular mobile service providers to carry intra-circle long distance traffic as well as by offering a similar policy regime as the FSPs with respect to interconnectivity, tenure of license, license fee in the form of revenue share, and so on. The policy also specified that DoT (and Mahanagar Telephone Nigam Limited, MTNL) would enter cellular mobile service provision as the third operator in each service area in a time-bound manner.

The stipulations in NTP 1999 were made operational by the government when it issued guidelines for issue of basic service licenses on January 25, 2001. These guidelines were issued after a consultation process with TRAI that saw the regulator give its recommendations of August 31, 2000, on issues pertaining to licenses for FSPs in 15 vacant circles and the migration of existing licensees to the revenue sharing regime, followed by a set of reconsidered recommendations given on October 31, 2000, and a final set of recommendations specific to limited mobility on January 8, 2001. The permission for limited mobility services was the watershed event in the years following the announcement of NTP 1999 and has had important effects on the competitive mix in the telecom services market that lasted into more recent times.

The nub of the controversy over limited mobility services arises from the categorization in NTP 1999 of telecom access services into cellular mobile service providers and fixed service providers. This division of access services among fixed and mobile was already being disrupted at that time by the emergence of technological fixes that utilized radio links to provide faster and cheaper telephone connectivity to telecom consumers. Traditionally, costs associated with connecting a subscriber's premises with the nearest telephone exchange were very high, and the process of wiring this 'local loop' was time-consuming.[14] Code Division Multiple Access (CDMA) technology, which was developed in the United States as an alternative to the mainly European Global System for Mobile (or, more correctly, Groupe Speciale Mobile) (GSM)[15] cellular technology, offered to solve the problem of cumbersome and costly wiring by enabling deployment of radio frequencies in the last mile of the fixed network as well. This was a game changer as far as FSPs were concerned, as they could deploy it to offer access through fixed wireless terminals as well as handheld telephones, which could be used within the entire 'local area' of the subscriber's telephone exchange.[16] The use of WLL was permitted – indeed, it was the preferred method for providing basic services – by provisions in the license agreement entered into with the first set of basic service providers, and this entailed that electromagnetic spectrum was to be made available to FSPs for this purpose. Even after WLL was deployed, the subscriber had only fixed phone access using fixed wireless terminals (FWTs). MTNL, which was the incumbent FSP in the metropolitan circles of Delhi and Bombay, was the first to propose the use of CDMA to offer what it termed cellular service with limited mobility in 1999. Notwithstanding MTNL's initial efforts to introduce limited mobility, it has been widely speculated[17] that the campaign to permit the service was led by one private sector entity – Reliance – that had not obtained many high revenue yielding circles in the two rounds of cellular mobile service tenders but was keen to enter the access services sector at that point in time. Irrespective of the veracity of this speculation, careful examination of the sequence of events that led to the executive permission for limited mobility service offering in the Indian telecom network reveals that the regulator actively contributed to bending the policy stipulation that compartmentalized fixed and cellular telephony into separate categories (and markets) by executive fiat. It could be argued that the policy was shortsighted in the first place for not recognizing the impending convergence in the access services market, but that begs the question as to what the appropriate course of action should have been for the policymaker and the role an independent regulator ought to have played in the process. The sequence of events that led to the unveiling of limited mobility services in India is another instance of the ambiguity that informs institutional relations between sector entities that treat the policy stipulation either as absolutely inviolate or merely as a broad guideline amenable to modification depending upon the balance of convenience rather than on any stated economic or administrative principle. Where a certain course of action (or inaction) is to be justified to the political economy, policy stipulations are drawn as red lines; if, however, they are inconvenient or – worse – prove to be unresponsive to exogenous developments usually occasioned by poor conceptualization in the first place, the same policy stipulations are jettisoned by using inventive justifications. A policy such as NTP 1999 does not prove to be a hindrance if the executive branch decides to observe elements of it in the breach, notwithstanding the presence of an independent regulator. In fact, it was the case with limited mobility that the regulator actively aided the executive decision-making process to facilitate its unveiling, primarily justifying it on the basis of consumer interest and common good while giving short shrift to contractual obligations. It is also of interest that the legislature acceded to the flexible interpretation of policy under the impression that "the development in technology permits new and an affordable facility to the consumers" (Standing Committee on Information Technology [SCIT] 2001a: 21)[18], and the higher judiciary (in the limited mobility litigation discussed in Chapter 3) did not adjudicate on the substantive dispute between CMSPs and FSPs that would have helped clarify the nature of the respective long-term contracts, but rather concentrated its efforts in the institutional direction of empowering the specialized tribunal's mandate. The chronology of the main events in the process is indicated in Table 4.2.

A cursory look at the chronology reveals the tight fit between institutional actors on the changing contours of policy on limited mobility. One way to consider the entire sequence of events is to view it as a convergence of opinion on the desirability of deploying new technology to provide cheaper service faster (some form of a policy network at work); such a view would, however, stretch credibility given the contestation more generally on a wide range of regulatory issues and specifically between interests representing the cellular licensees on the one hand and those representing FSPs on the other. The standard justification that the march of technology ought not to be hindered and

Table 4.2 Developments leading to limited mobility for fixed phone users and beyond

Date	*Event*	*Remarks*
November, 1994	Two cellular licenses each issued in four metros	GSM mandated (900 MHz band)
January 16, 1995	Bids for basic licenses invited	Along with bids for circle cellular licenses
December, 1995	34 circle CMTS licenses issued	In 18 circles; six vacant slots (two each in two circles and one each in two other circles); GSM mandated (900 MHz band)
September, 1997/March, 1998	First six 15-year basic licenses issued in six circles	Andhra Pradesh – Tata Teleservices; Gujarat – Reliance Telecom; Madhya Pradesh – Bharti Telenet; Maharashtra – Hughes Ispat; Punjab – Essar Commvision; Rajasthan – Telelink Network
October 10, 1997	MTNL CMTS license effective	Followed by litigation in Delhi High Court
October 7, 1998	DoT reference to TRAI for recommendations on license fee payable by circle CMTS licensees	Seeking recommendations on license fee payable in the extended period of license from 11 to 15 years
March 26, 1999/April 1, 1999	NTP 1999 announced/effective	Separate categories of FSPs and CMSPs
April 23, 1999	DoT reference to TRAI for recommendations on terms and conditions of license to FSPs under NTP 1999	Also for CMSP-related issues
July 12, 1999	DoT reference to TRAI for recommendations on license fee for migration of existing licensees to NTP 1999 regime	
August 1, 1999	Migration package to basic service licensees and CMTS licensees effective	20-year license tenure; no guarantee of duopoly to CMTS; provisional revenue share of 15%

April, 1999/August 23, 1999	DoT reported/MTNL sought TRAI's approval for tariffs for CDMA-based CMTS with limited mobility	Extensive TRAI-DoT correspondence initiated
September 1, 1999	MTNL changes service to "CMTS using CDMA technology"	References to WLL and limited mobility dropped
September 13, 1999	DoT permits CMTS using any technology, including CDMA	Technology neutral licensing
February 29, 2000	DoT (Department of Telecom Services, later BSNL) given CMTS license as third operator	21 circles and two metros (Chennai and Kolkata); MTNL also free to provide CMTS; GSM opted for (900 MHz band allotted)
June 23, 2000	TRAI's recommendations on CMSPs	Revenue share recommended to be 17% (10% A&N and J&K) of adjusted gross revenue (AGR)
August 31, 2000	TRAI's recommendations on FSPs	Revenue share recommended to be 12/10/8% for category A/B/C circles respectively; TRAI noted that employing WLL inescapable for quick rollout of fixed services
October 9, 2000	DoT reference to TRAI seeking reconsideration of FSP recommendations	Sought, inter alia, recommendations on allowing handheld terminals to FSP subscribers
October 31, 2000	TRAI's reconsidered recommendations on FSPs	Did not include limited mobility as it was a fresh issue
November 3, 2000	TRAI consultation process on limited mobility initiated	
December 29, 2000	DoT considered TRAI's reconsidered FSP recommendations	
January 5, 2001	Guidelines for fourth CMTS license issued	

(*Continued*)

Table 4.2 (Continued)

Date	*Event*	*Remarks*
January 8, 2001	TRAI's recommendations on limited mobility	Recommends WLL with mobility within short distance charging area (SDCA)
January 9, 2001/February 8, 2001	Parliamentary Standing Committee meets to consider limited mobility	Subject taken up for examination against the background of receipt of number of representations
January 22, 2001	*Petition no. 1* filed by COAI before TDSAT	Start of limited mobility litigation
January 25, 2001	DoT issues basic services guidelines	Licenses to be issued on 'continuous' basis; entry fee fixed at Rs 497 crore pan-India; 5 + 5 Mhz in 800 MHz band (CDMA) to be allocated for wireless operations in subscriber access network on first come first served basis
March 1/8, 2001	Parliamentary Standing Committee report adopted/presented to Parliament	Concludes that limited mobility is beneficial, subject to level playing field being ensured for operators
April 6/23, 2001	Decided to request/ notified the reference to the reconstituted GoT-IT to make recommendations on limited mobility before April 30, 2001	GoT-IT originally constituted on December 13, 1999, and reconstituted on November 10, 2000
April 26/27, 2001	GoT-IT recommendations on limited mobility made/accepted	
July–September, 2001	New FSP licenses issued	17 to Reliance, four each to Tata and Bharti
September–October, 2001	Fourth CMTS licenses issued for four metros and 13 circles (bids not invited for Assam, NE, J&K; bids not received for Orissa, A&N, Bihar, West Bengal)	Three-stage bidding with upfront entry fee (INR 1,633.57 crore); GSM opted for (1800 MHz band allotted); spectrum usage charges (SUC) of 2% of AGR for spectrum up to 4.4 MHz

March 25, 2002	TDSAT's first order on limited mobility dispute	Upheld limited mobility
December 17, 2002	Supreme Court order in *COAI* v. *UoI*	Remanded limited mobility matter to TDSAT
August 8, 2003	TDSAT's second order on limited mobility dispute	Upheld limited mobility by majority
October 27, 2003*	TRAI recommendations on unified licensing	UAS license recommended as first stage towards unified license (UL)
October 30, 2003*	Group of Ministers considered TRAI's unified licensing recommendations	
October 31, 2003*	Union Cabinet decision on course towards unified licensing	To enhance scope of NTP 1999 to include Unified Access Service Licence (UASL) and UL; TRAI recommendations on UASL accepted; DoT authorized to decide implementation details of UASL
November 11, 2003*	NTP 1999 amended	UL/UASL incorporated; migration procedure specified
November 14, 2003*	TRAI clarification that entry fee of new unified license would be the same as fourth CMTS license	
November 17/24, 2003*	DoT decision to accept and process UASL applications on similar lines as FSP licenses/issuance of letters of intent (LoIs)	first come first served (FCFS); continuous basis
January–May, 2004*	26 UAS licenses issued	

* These seven events occurred after limited mobility was a fait accompli. However, they are shown here as they belong to the longer sequence that culminated in a converged access services market, short of unified licensing.

Source: Compiled by author.

that CDMA offered a viable alternative to GSM is not as innocent as it appears at first blush given the Indian telecom sector context at that point in time. If achieving technology neutrality was the only objective, it would have sufficed for DoT to permit it *in the provision of cellular mobile service*, as it did when faced with the pre-amendment Act TRAI's queries on the scope of MTNL's cellular mobile service launch offering. Such a move – even if it could be criticized for expediency or conflict of interest – would still have meant that the NTP 1999 category of CMSP remained within the original contractual regime, albeit with modification of terms. However, by including limited mobility offerings by FSPs within the scope of technology neutrality, the contractual applecart was upset and the competitive matrix was modified irrevocably. The extraordinary consensus was clearly led and driven by the executive wing, and TRAI – in its post-amendment incarnation – helped this implicit mutation of policy such that the original stipulation in favour of GSM mobile technology[19] was relaxed to bring in limited mobility in the garb of introducing CDMA and technology neutrality. As noted earlier, limited mobility was the thin edge of the wedge that rapidly scaled up competition and inexorably altered the competitive scenario in the telecom sector. While that may arguably be a desirable objective, the process by which it was achieved suggests that either the policy formulation was poorly conceptualized in the first place or that policy was changed on the go to suit requirements without acknowledging the need to formalize the change by amending or adding to policy. It is not clear why the process followed for example at the time of approval of UAS licenses (later on in 2003) whereby an addendum was issued to NTP 1999 could not be followed at the time of permitting limited mobility also.

TRAI did not offer any credible explanation for its change of view regarding the permissibility of limited mobility under the then existing policy, even though it had – in 1999 – noted the specific policy stipulations for CMSPs and FSPs. It had in its pre-amendment avatar laid out three key NTP 1999 stipulations in the following manner: (1) NTP 1999, which provides for CMSPs and FSPs, did not mention any specific technology for CMSP; hence the technology to be used by CMSP is specified together with license conditions, which was GSM until September 13, 1999; (2) NTP 1999 mentions WLL in the section on FSPs; (3) any service called cellular mobile service with limited mobility was not part of NTP 1999 and "was settled with a response from DOT, dated 17 September, 1999" (TRAI 1999d: Annex II, III).[20] In its recommendations of January 8, 2001, the reconstituted TRAI acknowledged that "the License Agreement for the Basic

Service Operator provides for use of WLL, it does not permit mobility" (TRAI 2001a: 75). However, it justified its recommendation in favour of permitting limited mobility on the grounds that "it would be in the best interests of the consumer and the telecom services in the country" (ibid: 82). It then went on to set up the straw man of deciding whether limited mobility is the same as mobility offered by full service mobile telephone service and unsurprisingly concluded that it was not. It reflects on the techno-economic expertise of the regulator not to have noted that the core question was somewhat different: whether NTP 1999 contained a gap in categorization of access service providers that could – legitimately and without any derogation of existing terms of contract the licensor had entered into with different licenses – be covered without formally modifying the policy document. For all the clarity that NTP 1999 and the amending Act imparted to its role and functions, TRAI failed to address the crux of the limited mobility controversy: whether NTP 1999 allowed the provision of the service at all. It allowed many assertions in the DoT's reference to it go unchallenged from a regulatory standpoint, and the set of TRAI's recommendations leading up to the issuance of FSP guidelines in January 2001 reflects narrow technical adherence to the terms of reference rather than an examination of larger regulatory implications. DoT's change of preference can be viewed as political expediency at worst and as a manifestation of its clear conflict of interest between being licensor and service provider otherwise. It would, however, be difficult to find even weak arguments justifying TRAI's regulatory capitulation that was a result of its reactive approach compounded by a reduction in institutional stature as well as diminished self-perception.

The role of the political economy in nudging the decision towards permitting limited mobility can be discerned in the report of the GoT-IT[21] that also justified the move on the basis of the group's basic concern "to accelerate universal access to enable provision of telecom services at affordable rates specially in the rural areas" (GoT-IT 2001: 7). In answering its terms of reference, GoT-IT depended heavily on the January 8, 2001, recommendations of TRAI; the group also noted that TRAI's later recommendation "in favour of limited mobility was taken after following a transparent procedure of extensive consultation with all stakeholders and modifying terms applicable to cellular operators as stated above" (ibid: 5–6).

That limited mobility foreshadowed bigger changes in access service licensing is easy to discern in hindsight. Even when it came to be permitted as a policy measure, the convergence of access markets was rapidly gaining pace, and the half-way house solution of limited mobility received criticism from both sides. No change ought to have been made in the implementation of policy on such extraneous grounds in the first place, without even formally acknowledging that NTP 1999 required amendment. Even if the changes were unavoidable on the grounds of making available a more affordable service made possible due to technological change, a valuable opportunity for substantive reform of the policy framework was sacrificed at the altar of expediency. It would have made far more economic sense if the policy document had been drastically reworked to achieve true convergence of the access markets that had been artificially bifurcated in NTP 1999. By not even acknowledging that NTP 1999 required a modification at the time, DoT squandered the opportunity to correct the original mistake of mandating two separate markets by policy diktat. In not rising to its regulatory duty to closely examine all aspects of the proposed changes, TRAI was guilty of not only of a measure of policy circumvention by association, but of a more serious dereliction of the responsibility to set the regulatory agenda for the orderly growth of the telecom sector that is written into the very preamble of the statute that governs its functioning. That TDSAT upheld the permission granted to introduce limited mobility (by a majority, with the chair dissenting) is of little consequence since the dispute and its resolution arose from hasty implementation of faulty changes in the first place.

One final event of importance that came to a head in the closing year of the first phase needs to be recounted before closing this description. This development followed directly from the inability of the decision-making process to anticipate the extent of technological change that CDMA and WLL represented, coupled with some amount of regulatory laxity with respect to one entity possessing FSP licenses across many circles[22] and considered to be the main beneficiary of limited mobility permissions: Reliance. We have noted the intense litigation surrounding the introduction of limited mobility and the far-reaching influence of the Supreme Court's judgment in *Cellular Operators Association of India and Ors.* v. *Union of India and Ors.* (2003) 3 SCC 186. The second order of TDSAT in *Petition no. 1 of 2001* delivered on August 8, 2003, after the matter was remanded to the tribunal's attention by the apex court held in favour of the legality of the service by majority; it also required the government – in consultation with TRAI – to levy, "both on policy considerations as well as on economic grounds", an appropriate fee for permitting FSPs to provide limited mobility within the SDCA. Whether TDSAT could have gone beyond the strict terms of the dispute to fashion a decision that would have taken into account the rapid technological change that was the judgment's professed substantive justification is a moot question; the majority decision in any case offered suggestions on how to level the playing field that were admittedly skewed by the executive decision to permit limited mobility to FSPs. TRAI was less circumspect in dealing with the core issue raised in the limited mobility litigation that was soon to be placed at its doorstep: while it released a consultation paper on August 23, 2003, to follow up on and implement TDSAT's majority decision (after a formal reference was received from the government), it had also – perhaps anticipating the developments to occur in the immediate future once the tribunal's judgment was delivered – already issued a consultation paper on July 16, 2003, on issues pertaining to unified licensing for basic and cellular services that – if belatedly – discovered and actively propagated the virtues of convergence of wireless and wireline technologies and markets. This earlier consultation paper (TRAI 2003b) was part of the effort to sidestep the contentions emanating from the regulatory decision in 2001, at which point convergence was an imminent game changer for the sector but was ignored in the rush to permit limited mobility to FSPs. In the process, the consultation paper sought to project limited mobility as a *restrictive* license condition that militated against technological progress and acted as a regulatory constraint rather than as a modification in the terms of the larger contract that *facilitated* one set of licensees (FSPs) in providing service features they were prohibited from providing in the first instance because that would have adversely affected the contractual rights of another set of licensees (CMSPs). This flipping of what was by general consensus a concession given to FSPs into an apparently rigid regulatory constraint inimical to innovation and technological evolution cleverly catapulted TRAI from the horns of a regulatory dilemma to the high ground of techno-economic expertise from where it could champion the cause of convergence. Indeed, the regulator's recommendations of October 27, 2003, emanating from this consultation expressed the view that "licensing restrictions should not come in the way of technological developments as the artificial restrictions would encourage service providers to find loopholes in [the] licensing regime and they will use technology or loopholes in networks/regulation to by-pass such restrictions" (TRAI 2003c: 2).

This elaborate exercise meant to ostensibly usher in unified licensing could not mask one crucial shortcoming in the recommendations. What TRAI rather grandiosely labelled as recommendations on unified licensing in effect turned out to be a roadmap for unifying the access services market alone, without considering the unification of long distance services. A two-stage process was recommended – first an immediate migration[23] to a UAS license and a second and final movement to a 'fully unified' license at an indeterminate future date –[24] and was justified on the grounds of achieving rapid growth of teledensity, which TRAI viewed as the primary and immediate national objective that "revolve[d] around access network and the need to make available low cost access" (ibid: 19). Apart from being a circular argument that downplayed the role and effect of the long distance markets in determining access service prices, this approach of TRAI wasted another opportunity to effectively reset market dynamics to achieve convergence. The ad hoc recommendation to introduce UAS licensing whereby "service providers may offer basic and/or cellular services using any technology" (ibid) gives impetus to the view that this step was actually the next stage in finding FSPs a route into cellular mobile access service territory as initiated with the introduction of limited mobility, rather than an earnest effort to usher in convergence. This view is strengthened by the way in which TRAI dealt with the brazen breach of the terms of the limited mobility contract by Reliance Infocomm, an FSP licensee.

That the terms of the amended FSP license were breached by Reliance was never seriously disputed. The basic service guidelines of January 25, 2001, stipulated that limited mobility was permissible within the SDCA and that handover of calls from one SDCA to the next was prohibited. In its recommendations of October 27, 2003, TRAI noted that Reliance Infocomm – a basic service operator – competed right from the effective date of its license agreement as a cellular mobile service provider (except that the service drops at the time of moving from one SDCA to another) "by obtaining a license as a BSO [basic service operator] almost throughout the country and using multiple registration/call forwarding facility" (ibid: 23). TRAI did not treat this as a breach of license condition that would have entailed commensurate penal consequences. Instead, it chose to treat the offending licensee "at par with fourth cellular operator right from the date of signing their existing BSO license agreement" with the effect that it was seen as being in breach of conditions stipulating payment of appropriate entry fee paid by the fourth CMSPs in the respective circles along with penal interest as applicable to CMSPs. This treatment of a clear breach of contract conditions by the licensee as a compoundable offence and monetizing it ignored the spirit if not the letter of section 11 (1) (a) (3) of the TRAI Act, which requires the regulator to make recommendations on the matter of "revocation of licence for non-compliance of terms and conditions of licence". What is at dispute is not whether TRAI necessarily had to recommend revocation of the offending service provider's license; it could still have made a case for other, less severe, penalties. That TRAI did not even consider the flouting of the stipulation in the guidelines as a breach of contract worth further regulatory scrutiny is what attracts critical attention to the regulator's resolution of the matter. The summary treatment of the episode as a competition issue that occasioned only a monetary penalty limited to the penal interest on the entry fee payable as the fourth CMSP ignored other options and license conditions that could have been recommended by TRAI for invocation by DoT to ensure that the breach of contract conditions was appropriately addressed. In its recommendations of October 27, 2003, TRAI noted that other FSPs "have not breached the condition either in letter or spirit and have confined mobility within SDCA" (ibid: 24), but the only difference in treatment of these licensees and Reliance Infocomm was the additional penal interest amount that the FSP breaching the condition (both in letter and spirit) was required to pay. The TDSAT's minority judgment in *Petition no. 1 of 2001* had observed of to the decision to permit limited mobility that it appeared that "the considerations which led the Government to pass the impugned decision were not the result of bona fide exercise of power and were motivated by extraneous considerations to confer undue benefit on the FSPs". TRAI's recommendation (agreed to and acted upon by the government) to treat a major breach of license condition by one FSP as a minor infraction liable only to monetary penalty is another contentious example of the uneven quality of regulatory decision-making on crucial issues related to the access services market.

The cost to be incurred by FSPs for migration to a UAS license that would allow them to offer cellular mobile services was the difference between the generally higher entry fee paid by the fourth CMSP in the 2001 bidding round and the entry fee paid by the FSP for the respective circle at the time of their own entry (see Table 4.3 and Table 4.4). One FSP admittedly in breach of license conditions was visited with a monetary penalty that constituted less than half of the entry fee it was required to pay for being allowed to provide full mobility to its subscribers. It is rather obvious that the penal pain – a mere rap on the knuckles – was not an unbearable burden given the benefit of providing cellular services across a large swathe of the Indian market just warming to the pleasures of handheld telephone access. The recommendation of TRAI signalled regulatory reluctance to enforce contractual commitments and represents a diminution of the credible commitment the presence of an independent regulator engenders in the economics of the sector.

Table 4.3 Entry fee paid by FSPs and fourth CMSPs in 2001

(Entry fee in INR crore; 1 crore = 10 million)

Category	*Circle*	*FSP entry fee prescribed by January 2001 guidelines*	*Entry fee bid by fourth CMSP in 2001*	*Remarks*
A	Andhra Pradesh	35.00	103.01	
	Delhi	50.00	170.70	
	Gujarat	40.00	109.01	
	Karnataka	35.00	206.83	
	Maharashtra	115.00	189.00 + 203.66 (Mumbai)	Included Mumbai and Goa for FSP
	Tamil Nadu	50.00	79.00 + 154.00 (Chennai)	Included Chennai for FSP
B	Haryana	10.00	21.46	
	Kerala	20.00	40.54	
	Madhya Pradesh	20.00	17.45	Included Chhattisgarh
	Punjab	20.00	151.75	
	Rajasthan	20.00	32.25	
	Uttar Pradesh (West)	15.00	30.55	Included Uttaranchal
	Uttar Pradesh (East)	15.00	45.25	
	West Bengal	25.00	0.00 + 78.01 (Kolkata)	Included Kolkata for FSP; no bid received for WB CMSP
C	Andaman & Nicobar	1.00	0.00	No bid received for CMSP
	Assam	5.00	0.00	No bid called for CMSP
	Bihar	10.00	0.00	Included Jharkhand; no bid received for CMSP
	Himachal Pradesh	2.00	1.10	
	Jammu & Kashmir	2.00	0.00	No bid called for CMSP
	North East	2.00	0.00	No bid called for CMSP
	Orissa	5.00	0.00	No bid received for CMSP
	Total	**497.00**	**1,633.57**	

Source: TRAI 2003c: 40–41, DoT.

Table 4.4 FSP migration to UAS licensing

FSP name	*No. of circles*	*Difference between entry fees (INR billion)*	*Penal interest (INR billion)*	*Total (INR billion)*
Reliance	18	10.96	4.85	15.81
Tata	6	5.45	–	5.45
Bharti	5	4.87	–	4.87
Shyam	1	0.296	–	0.296
HFCL*	1	–	–	–

* HFCL paid a higher FSP entry fee than the fourth CMSP entry fee in Punjab.
Source: TRAI 2003c: 42.

This example vividly captures the nature of regulatory decision-making in the face of conflicting group interests in a market characterized by rapid technological change and where long-term incomplete contracts were just beginning to gain traction. Where the early movers in a competitive access market gained advantages due to changes in technology that made cellular mobile access the preferred mode of telephony, the followers and late entrants who had initially placed their bets on a less attractive technological option (as fixed access service turned out to be) lobbied the system and managed to obtain the crucial competitive breakthrough by obtaining regulatory permission to provide limited mobility in their networks. One entity in particular used the window of opportunity that limited mobility provided to tread on the wrong side of the law to give unexpectedly stiff competition to incumbent CMSPs. Rather than penalizing these breaches of contract conditions, the regulatory regime changed the rules of the game to allow FSPs to migrate to a license that permitted 'full mobility' to be offered on a technologically neutral platform, justifying the decision on the basis of technological change and the need to foster innovation rather than stifle it. In terms of institutional analysis, it is necessary to note that while limited mobility was permitted at the behest of the executive branch with the regulator playing a secondary – and somewhat grudging – role, the migration offer to UAS licensing saw the regulator taking the lead and initiating consultation *suo motu* even as the matter was being litigated before the specialized tribunal, and the government's role was limited to seeking TRAI's recommendations pursuant to TDSAT's judgment on limited mobility that was ultimately reduced to a redundant exercise in view of the UAS licensing recommendations. It is interesting that the person at the helm of TRAI had changed from the limited mobility decision to the UAS licensing decision; however, it would be naïve to attribute a "resolution of the WLL crisis" that "boosted investor confidence and gave birth to India's telecom revolution" to the latter chairperson on the strength of the UAS licensing decision (Mukherji 2009: 512, footnote omitted). That would only serve to exalt expediency and read regulatory percipience into what was essentially acquiescence to the completion of the project initiated earlier by the executive branch of shepherding late entrant FSPs into the more lucrative cellular access service market.

By the same token, it would be farfetched to view the unified access licensing recommendations (including the decision not to penalize one FSP for breach of license conditions) as a "solution" that "compromise[d] between strict legal principles and the need for technological innovation" (Raghavan 2007: 423). In such a view, the permission to introduce limited mobility by the January 25, 2001, guidelines would equate to a 'strict legal principle', which begs the question of whether such a move was legally valid in the first place, notwithstanding the majority decision of TDSAT that attained finality only because of a later compromise brokered between the FSPs and the CMSPs on the back of other concessions that resulted in the appeal being withdrawn from the Supreme Court. Even if the decision is considered legally valid on the basis of the endorsement given by the tribunal's decision, it still fell short of addressing the imperatives of convergence clearly discernible at the time of the introduction of limited mobility itself, if not earlier when NTP 1999 was drafted and approved. The flaw in the policymaking effectuated in 1999 and perpetuated in 2001 cannot be hailed simultaneously as a legal principle and as a trigger for change and technological innovation. While the non-penalty imposed on Reliance Infocomm for breach of license conditions is perhaps weak evidence to conclude that the entire process (of FSP with only fixed access first morphing to FSP with limited mobility and then to UAS licensee authorized to provide mobile access) was tweaked to suit the interests of one group of service providers to the detriment of others, it is nevertheless evidence enough that powerful interest groups influenced the course of regulatory decision-making in the Indian telecom sector with reasonable success during this period.

This is not to say that the countervailing interests of CMSPs got regulatory or policy short shrift. The early entrants to the competitive access service market did indeed fiercely resist the FSPs' moves to gain a foothold in their market. In fact, CMSPs fashioned a multi-pronged strategy to protect their turf; while simultaneously pursuing the legal option right from the inception of limited mobility services as a means of protecting their property rights and delaying the entry of further competition, they also lobbied the executive and regulatory arms of the Indian state to gather as many concessions as they could in the event that competition from FSP licensees – some of whom were large multi-industry conglomerates – eventually became a reality in the mobile telephone services market. CMSPs – in tandem with FSPs – had tasted success already in their earlier efforts to wrangle concessions from the state in the form of the migration package in 1999 that allowed them to withdraw from their contractual license fee commitments and substitute them with a revenue sharing regime. The lessons from this prior experience and its positive outcome stood the CMSPs in good stead in their second effort in the period between 2001 and 2003. The coinage 'level playing field' gained considerable currency during this time as CMSPs battled the expansion of the service provisioning remit of FSPs by insisting that the policy interpretation that allowed FSPs to provide mobile telephone services required matching financial and non-financial concessions the CMSPs deserved in lieu of giving up their exclusive contractual right to provide such services. The second TDSAT order (majority decision) of August 8, 2003, in the limited mobility litigation formally endorsed the CMSPs' plea by agreeing that they had not been fully compensated for the rollout of limited mobility services and suggesting measures to restore a level playing field. What is of greater interest, however, is the following listing of concessions given to CMSPs that deserves quotation in extenso:

> Keeping in view the impact of WLL (M) service on the operation of cellular operators as also the recommendations of TRAI in this regard, the Government reduced the licence fee in respect of cellular operators from 17% to 12% for metros and Category 'A' circles, 10% for Category 'B' circles and 8% for Category 'C' circles and brought the same on par with annual licence fee required to be paid by the basic service operators resulting in substantial benefit in monetary terms to Cellular Operators. They were also allowed reimbursement of 5% of access charges. Other reliefs given included permitting cellular operators to also deploy fixed telephone on their GSM networks, levying of same frequency spectrum charges for both the cellular operators and basic service operators, making available increased interconnection point at TAX-I and TAX-II levels to cell operators and grant of permission to cellular operators to establish and operate mobile PCOs within their service area.
>
> (Paragraph 63, second TDSAT order of 8.8.2003)

TRAI's recommendations of October 27, 2003, put a monetary value on some of these concessions. The recommendations estimated that (1) the value in Net Present Value (NPV) terms that CMSPs derived from the migration package to the revenue sharing regime up to 2003–4 was INR 4,565 crore (TRAI 2003c: 26);[25] and that (2) the reduction of license fee in January 2001, while permitting limited mobility from 15 per cent to 12, 10, and 8 per cent respectively for category A, B, and C circles, resulted in a relief of INR 15,015 crore based on NPV during the validity period of the license (ibid: 26–27). Based on these numbers and the other concessions already given, the recommendations concluded that no further concessions were due to CMSPs and that the aspect of level playing field mentioned in the Supreme Court's verdict in the limited mobility litigation "had been taken care [of] in these recommendations" (ibid: 27). The executive branch went one better on this recommendation, however: it subsequently reduced the license fee by 2 per cent in each of the categories, making it 10, 8, and 6 per cent of the adjusted gross revenue for metro/category A, category B, and category C circles respectively with effect from April 1, 2004. In addition, the first two cellular licensees in each circle were given an additional reduction of another 2 per cent for a period of four years with effect from the same date (subject to their paying a minimum of 5 per cent, which was the stipulated USO levy) ostensibly to further level the playing field for these early entrants.

It is safe to assume that the all-round munificence resulting from executive kindheartedness and regulatory laxity in the limited mobility issue was even-handed enough to induce the truce that persuaded CMSPs to withdraw their appeal against the second TDSAT decision before the Supreme Court. It is nonetheless true that the cessation of hostilities was brought about at considerable regulatory cost and the controversy and charges of partisanship that attended the resolution of the largely executive-made crisis (aided in part by regulatory sloth) continued to affect stakeholder perceptions of regulatory efficacy and policy continuity in the sector for a long time. The antagonisms of the first stage of telecom regulation in India festered well into the next two stages and show no signs of yielding to a permanent peace even after the lapse of a decade. The seeds of the suspicion that policy and regulatory support could be manipulated for individual and group interests irrespective of the form and content of contracts among stakeholders were sown in this period, and the Indian telecom sector has not yet recovered from the aftermath of the bitterness engendered by the half victories wrangled by the principal protagonists at considerable cost to their balance sheets as well as to their corporate reputations.

Notes

1 This phase division broadly corresponds with TRAI's own perception as well (see TRAI 2005d; TRAI 2012c).

2 'Affordability' was also seen as a requirement. In fact, TRAI sought to achieve the golden mean between "several conflicting requirements" (TRAI 1999a: 11) without specifying what these were. The primary focus was to undertake inflation adjustment of costs in each part of the telecom network as a first step towards a fully balanced and cost adjusted tariff structure.

3 The looming presence of landlines had an interesting fallout on participation: in TRAI's tariff-setting exercise, "an overwhelming majority" of comments on the regulator's proposals for cellular tariffs came from service providers and the Cellular Operators Association of India (COAI), while subscribers and consumer associations "responded mainly to the proposals for Basic Services" (TTO 1999: 84), reflecting a basic regulatory conundrum. This is further evidence that participation – even wide and varied– is no guarantor of interest balancing or balanced regulation.

4 TTO 1999 noted that DoT's annual surplus in 1997–98 was about Rs 7,000 crore (INR 70 billion, or about USD 1.88 billion at the prevalent average exchange rate for that year; see Dua and Ranjan 2010: 23) while DoT's projected total revenues for 1999–2000 were about Rs 23,181 crore (INR 232 billion/USD 5.35 billion). It also expressed the view that it was "necessary for the DOT to have substantial surpluses for expanding the network and to meet other social obligations" (TTO 1999: 54).

5 The foreign participants who showed interest included AT&T, US West, Bell Atlantic, Nynex, Swiss Telecom, NTT, Bell Canada, Bezeq, and Shinawatra (Dokeniya 1999).

6 The licensees were allowed to carry intra-circle traffic on their networks.

7 Note that two cellular licenses were on offer in each circle compared to only one for basic services. Even on this basis, only 6 out of 21 circles saw basic service licenses being made effective (against 13 letters of intent issued by DoT) compared to 18 (out of 20 circles for which tenders were called) in the case of cellular licenses.

8 Introduction of wireless in the local loop was a valid competitive response as well since it drastically reduced capital costs for the basic service licensees.

9 Industry lobbies also got together to form a group on November 26, 1998, that was referred to as InGoT "to provide a co-ordinated response to the government's GoT" (Uppal et al. 2006: 21).

10 The Joint Parliamentary Committee (JPC) in its report in October 2013 indicates that a review meeting chaired by the prime minister on November 1, 1997, decided that a study on the health of the cellular industry "might be referred to [TRAI] which had an advisory role under the TRAI Act" (JPC 2013: 6). One opinion offered to the committee for DoT finally not referring the matter to TRAI was that "TRAI was at its nascent stage" and that it was "busy with re-balancing and restructuring of tariff for the entire telecom sector which was a fundamental exercise" (ibid: 7). The JPC concludes that DoT lost "the benefit of input from [TRAI] at a critical juncture" (ibid: 148).

11 See, for example, "New Tariff on Hold to Keep Mamata in Fold", *Indian Express*, March 11, 1999.

12 These consultations were held at the official level (i.e., not by members of the authority as constituted by the TRAI Act but by officers working in the secretariat) according to the first amendment to TTO 1999.

13 The political economy also had a role in DoT's inability to follow the logic of tariff rebalancing. In view of the impracticality of increasing rentals and local call charges to compensate for the steep decline in long distance surpluses even in a phased manner, DoT sought to compensate BSNL (and other fixed service providers) by including the component of 'access deficit charges' (ADC) in the interconnection usage charges (IUC) payable by interconnecting networks to fixed service providers in a TRAI regulation (no. 1 of 2003) issued on January 24, 2003 (TRAI 2003a). ADC was provided to cover shortfall in revenues for access services that were being provided below cost. For a critique followed by a rebuttal in discussion by a former TRAI official who issued the IUC regulation, see Noll and Wallsten 2007. ADC proved to be a mixed blessing for BSNL, given that its withdrawal by TRAI in 2008–9 was followed by BSNL posting its first full year of loss in 2009–10.

14 Sridhar estimates that the cost of connecting a consumer to the first point of switching in the telephone network "is close to two-thirds of the cost of the network" (2012: 42).

15 CMSPs were obliged under the terms of their license to offer services using GSM technology.

16 Recall that SDCA was coterminous with the local area for charging for trunk calls. It corresponded to the administrative division ('taluka'), was allotted a unique trunk code, and calls within it required dialing the subscriber number without prefixing the long distance access code ('0') (see NNP 2003).

17 Uppal et al., for example, observe that "private basic businesses had failed to take off in any significant manner and mobility offered a major opportunity to companies like Reliance who had acquired basic licenses for almost the whole country post NTP-99" (2006: 9). It has also been noted that "the notable beneficiary of the WLLM policy was Reliance Infocomm (with licenses in seventeen circles)" (Mukherji 2009: 507).

18 The Parliamentary Standing Committee on Information Technology circumscribed this view by noting that, subject to necessary policy changes, WLL would "greatly benefit the consumers provided the system, as envisaged, operat[es] fairly in the interest of consumers" (SCIT 2001a: 24).

19 This was not explicit in NTP 1994 but was a feature of the invitation of bids in the first rounds of cellular licensing. NTP 1999 also did not explicitly mention either technology, but the addendum issued to it in November 2003 permitted use of 'any technology'.

20 What TRAI left incomplete was whether the DoT letter could be construed to imply that FSPs were also barred from providing WLL-mobile services.

21 Other than the finance minister (chair), and ministers of communications, information technology, information and broadcasting, and law, justice and company affairs, GoT-IT had as members eminent jurist Fali Nariman, MP, the central vigilance commissioner (N. Vittal, who had been

secretary, DoT); M S Ahluwalia, member, Planning Commission; and Sudheendra Kulkarni, director in the Prime Minister's Office. Secretary, DoT was the member-secretary.

22 Recall that Reliance had acquired 17 FSP licenses in 2001 in addition to the Gujarat FSP license it won in 1997 in the first round. It signed the license agreements for Rajasthan, Uttar Pradesh (East), Karnataka, Punjab, Andhra Pradesh, Haryana, Kerala, Uttar Pradesh (West), West Bengal, Madhya Pradesh, Bihar, Himachal Pradesh, Orissa, Tamil Nadu, Delhi, and Andaman & Nicobar circles (and Maharashtra) in July–September 2001, paying a total entry fee of INR 410 crore (INR 4.1 billion) (TRAI 2003c: 42).

23 TRAI recommended introducing UAS licenses with an option given to existing licensees to migrate to it, strictly speaking. However, it was even then rather obvious that the option *not* to migrate would realistically be utilized only by CMSPs; there was little chance that FSPs would not grab the option to migrate given the clear advantages of cellular access provision over fixed access.

24 It must be conceded that TRAI recommended that the second stage of unified access licensing through authorization (fully unified licensing) "should be initiated" for all services and all geographical areas (TRAI 2003c: 31). In reality, the second stage did not achieve fruition until 2013, after NTP 2012 was unveiled, although TRAI did make another set of follow up recommendations on unified licensing on January 13, 2005 (TRAI 2005a) that were not effectuated.

25 There is no reference to the concomitant benefit that accrued to the six FSPs that were also beneficiaries of the migration package, a point that the comptroller and auditor heneral had in fact objected to as grossly unjustified and premature (see CAG 2000).

5

TRANSITION TO UNIFIED ACCESS

The imprint of NTP 1999 is writ large on the fortunes of the telecom sector during this phase since it significantly altered the framework of sector governance to the advantage of service providers. Table 5.1 captures the telecom sector's competitive snapshot at the beginning of the second stage.

Some other developments are of interest to the narrative. The first is the corporatization of the incumbent government monopoly into Bharat Sanchar Nigam Limited (BSNL) when it was incorporated on September 15, 2000.[1] This again followed the promise made in NTP 1999 to corporatize the Department of Telecommunications (DoT) without, however, indicating any intention to privatize its operations. BSNL provided access service all over the country (except in Delhi and Bombay where another government corporation – Mahanagar Telephone Nigam Limited (MTNL) – had already been in operation since 1986) as well as acted as the incumbent service provider in the domestic long distance market that NTP 1999 had opened for competition on January 1, 2000, in line with the country's World Trade Organization (WTO) commitments.[2] BSNL became the third CMSP in October 2002 when its services were launched, even though it held licenses dating back to February 2000. The entry of competition in domestic long distance had seriously dented its ability to cross-subsidize its local access market,[3] but BSNL did manage to rapidly increase its market share in the growing cellular market in the initial stages.[4] The presence of – and relatively strong response to competition from – the government-owned company had implications for regulatory performance during this period. However, the fact of corporatization itself, along with the intense competition engendered by the policy framework that increased the number and heft of private sector stakeholders in the regulatory space, gradually led to a rebalancing of institutional authority since DoT no longer combined the service provisioning and

Table 5.1 License types in operation by 2004

	License type	*Year of entry*	*Remarks*
1	Incumbent fixed telephone service	–	Incumbent public telecom operator; no explicit license
2	Cellular mobile telephone service (CMTS, metro)	1994	Two in each of four metros; Group Speciale Mobile (GSM) 900 MHz
3	Cellular mobile telephone service (circle)	1995	Two in each of 18 circles; GSM 900 MHz
4	Basic telephone service (first round)	1997/1998	One each in six circles
5	Incumbent CMTS	1997	Mahanagar Telephone Nigam Limited (MTNL); license stayed by court
6	Incumbent CMTS (third operator)	2000	DTS (Bharat Sanchar Nigam Limited, BSNL)/ MTNL; GSM (voluntary) 900 MHz
7	Fixed service provider (second round)	2001	With limited mobility; Code Division Multiple Access (CDMA) 800 MHz
8	Cellular mobile service provider (fourth operator)	2001	One each in 4 metros/13 circles; GSM (voluntary) 1800 MHz
9	Unified access service	2004	Migration permitted; GSM 1800 MHz, CDMA 800 MHz

Source: Compiled by author.

policymaking roles within itself. The separation of the policy-making role of DoT from its service provisioning function had the effect of reducing the ability of each of the parts to act in tandem to fluster the regulator. It can also be argued that full government ownership of BSNL worked to stifle rather than sustain it in taking advantage of regulatory developments in the long run. Corporatization did not alter the organizational culture of the new entity, and the lack of operational freedom coupled with soft budget constraints portended all-round decline in a potentially intensely competitive market. At the beginning of 2004, however, BSNL was still sanguine about its chances as a full service operator.

The corporatization of BSNL, the entry of a number of private sector service providers in the access service markets, and the introduction of competition in the long distance markets reinforced the Telecom

Regulatory Authority of India (TRAI)'s tariff-setting role in no small measure around this time. TRAI had always been intended to be the final authority on tariff related issues under section 11 (2) of the TRAI Act. However, tariffs to consumers of services crucially depended upon the nature of payment arrangements between service providers whose networks were interconnected. Towards this end, the Amendment Act in 2000 had introduced section 11 (1) (b) (ii) that made it a function of TRAI to "fix the terms and conditions of inter-connectivity between the service providers" notwithstanding anything contained in the terms and conditions of the license. This function was in addition to the previously existing mandate to "ensure technical compatibility and effective interconnection between different service providers" (section 11 (1) (b) (iii)) and to "regulate arrangement among service providers of sharing their revenue derived from providing telecommunication services" (section 11 (1) (b) (iv)) that were carried over from the pre-amended statute. The complexity of interconnection arrangements in a multi-operator environment demanded a professional response from TRAI for regulating the charges payable. The Telecommunication Interconnection Usage Charges (IUC) Regulation, 2003 (TRAI 2003a), that was issued by TRAI on January 24, 2003, under section 36 of the TRAI Act, using its powers under clauses (ii), (iii), and (iv) of section 11 (1) (b),[5] was its preliminary attempt to substantially alter the nature of interconnection payment arrangements put in place by Telecommunication Interconnection Regulation (TIR) 1999 (TRAI 1999c). While this first IUC regulation was superseded by the Telecommunication Interconnection Usage Charges Regulation 2003 (TRAI 2003d) issued on October 29, 2003, its importance lies not only in setting out the cost-based IUC regime in terms of origination, transit, and termination of calls in a multi-operator environment, but also in laying down that origination and termination charges would include access deficit charge (ADC) payable to fixed service providers (FSPs) "which they must get in order to keep the rental as well as local calls affordable" (TRAI 2003a: 15). ADC was a result of the regulator's tariff rebalancing efforts carried out earlier that led to steep declines in long distance tariffs calling for a different mechanism to ensure that below-cost retail rental and local call tariffs continued to be maintained in the fixed networks "to promote both Universal service and Universal access as per NTP '99" (ibid: 16). The first IUC Regulation is also important for a second reason: it noted that payment of termination charges becomes inescapable "irrespective of the type of network in which the call terminates" (ibid: 26) and proceeded to introduce a mobile termination charge (MTC) for calls terminating in a CMSP's network on the same principle as for FSPs, thus effectively introducing a calling party pays (CPP) regime and eliminating the need for cellular subscribers to pay airtime charges for receiving calls. While ADC has been subjected to withering criticism and much lobbying by CMSPs and ILD service providers for being inefficient and has been phased out over time to be entirely replaced by a lump sum charge in the form of a USO levy, the CPP regime introduced by the regulator in this indirect manner (without its again being subjected to the legal test as in the first attempt in 1999) endures in the Indian telecom sector.

That the resources for meeting USO would be raised through a universal access levy that was to be a percentage of revenue earned by licensed service providers was also specified in NTP 1999. NTP 1999's universal service commitment was to "provide access to all people for basic telecom services at affordable and reasonable prices". The broad mechanism by which USO was to be implemented entailed that FSPs would undertake access provision in rural and remote areas and claim reimbursement of costs from the funds collected by imposing the levy. The percentage of revenue share was to be decided in consultation with TRAI; in fact, DoT went farther than the NTP 1999 mandate and sought the regulator's recommendations in two tranches on May 21, 1999, and October 13, 1999, on (1) what class of operators would fund the levy and (2) the various possible cost models and approaches to determine the percentage contribution from operators' revenue and the computation mechanism along with per unit subsidies for village public telephones and direct exchange lines separately to cover capital and recurring expenditure and so forth (TRAI 2001b).

TRAI recommended on October 3, 2001, that the universal service levy (USL) should be levied at the rate of 5 per cent of the adjusted gross revenue (AGR) of all categories of network operators and carriers (such as FSPs, CMSPs, and national long distance [NLD] and international long distance [ILD] operators) other than pure value-added service providers such as internet service providers (ISPs) who did not own facilities. USL was to continue as a part of the license fee.[6] TRAI also recommended that USO implementation be clearly divided into two streams – stream I for the provision of common access and stream II for providing household telephones in high net cost areas. While these recommendations found broad acceptance in DoT, what did not was the regulator's recommendation that an independent unit to administer the USO fund may "be established by TRAI in consultation with the Government" (ibid: 39). Instead, while a universal service support policy came into effect from April 1, 2002 (a date also recommended by TRAI), a non-lapsable USO fund was established under the auspices of the central government by virtue of the Indian Telegraph (Amendment) Act, 2003, to amend the Indian Telegraph Act (ITA) and inserting Part IIA in it with retrospective effect. The office of the USO administrator was thus created as an attached office within DoT[7] and tasked with the duty under ITA to fulfil the government's universal service obligations. TRAI had also proposed in its recommendations the development of proxy cost models to assess the optimal cost of public and household telephones so as to match subsidies with costs,[8] but this recommendation too did not find favour with DoT, which opted for a bidding process whereby operators who quoted the least subsidy and provided services in targeted rural and remote areas would win the contract. As it turned out, most of the disbursements from the USO fund accrued to BSNL,[9] attracting the criticism that the country's universal service policies deter investments in areas they hope to target and that "subsidies discourage competition, and the most efficient operators are taxed to support the least efficient operator" (Noll and Wallsten 2006: 274).

Spectrum: scarcely understood resource

One last aspect needs attention before finally moving on to the second stage: a brief overview of the evolution of the spectrum management regime in the Indian telecom sector. Spectrum is often described as a scarce resource; in the Indian telecom sector's context, it was also a scarcely understood resource, and we contend that the roots of the disorderly treatment of property rights in it are embedded in the ambiguities that attended the design of the initial conditions of privatization of access service provision in the early 1990s. Spectrum was not central to the privatization process in its early stages, and the focus on de-monopolization and parting with the government's exclusive privilege under ITA of "establishing, maintaining and working telegraphs", including the bitter political contestation surrounding that process, so altered the terms of regulatory discourse that this scarce resource was not accorded its deserved centrality in the design of policy and auction methodology. In the initial rounds of tendering for cellular access service provision in the metros and circles for the first and second CMSPs, it was *not* usage rights in electromagnetic spectrum that were being auctioned; what was at stake was the right to establish and run telegraph service of a particular kind, and this was what the bidders tendered for. The first round of licensing was clearly not intended to be "a mechanism for capturing the value of the spectrum" (Sridhar 2012: 78), as even astute observers of the Indian telecom scene have noted, but actually a method to identify in a rather ad hoc fashion potential partners from the private sector interested in providing 'value-added services' that would supplement DoT's efforts to increase fixed phone teledensity, as already noted in an earlier section. Spectrum was not a key element in the decision mix at this time, and proof of its relative insignificance is available in the related tender documents. The foundational ambiguity regarding the criticality of spectrum to the whole enterprise of cellular access continued well into the next round of auction for the fourth CMSP in 2001 as well, although the decision-making process at this time benefited from previous experience enough for it to recognize that spectrum commanded a value in the marketplace.

The Amendment Act, while including the "efficient management of available spectrum" under section 11 (1) (a) (viii), did not backstop the recommendatory function with any stipulation making it mandatory for the government to seek its recommendations. Perhaps the first opportunity for the regulator to apply itself to the task of spectrum management came when it received the references from DoT post-NTP 1999 seeking its recommendations on license fee percentages in the revenue sharing regime and the further opening up of the basic and cellular services markets. TRAI's recommendations on issues pertaining to CMSPs were made in June 2000 and those on FSPs in August and October 2000. These two sets of recommendations are conspicuous for the contingent nature of TRAI's views on the subject of introduction of further competition and the timidity on display in the reluctance to encounter spectrum-related issues head on in order to reduce the level of contingency and ad hoc–ism in the recommendations.

The new spectrum management framework palpably reflected the legacy of the previous decade that did not care to attempt the valuation of this scarce resource, allowing its price to reveal its inherent worth as a key input in mobile telephony. The regulatory recommendations of 2000 and 2001, while they brought additional competition to the sector, also achieved policy lock-in of the administratively determined usage charge regime and rendered the path to market-based determination of value more difficult at a later stage. As could be expected, this regulatory view of spectrum pricing created a path-dependency entailing other policy and regulatory action leading to further distortions such as subscriber linked criteria (SLC) for allocation of additional spectrum and introduction of dual-technology licenses in the stages to follow. The benign interpretation of TRAI's regulatory stance is that the regulator was interested in pushing the policy objective of affordable telecom services at the cost of market determination of the scarcity value of spectrum. Notwithstanding that such a view foregrounds its own set of issues (since price is in fact a proxy for quality and TRAI's own experience with legacy issues in tariff policy pointed clearly to a trade-off between quality of service [QoS] and subsidized prices), this interpretation is not supported by evidence of any deliberation on TRAI's part on this aspect, led alone explicit articulation of any such logic. It has been theorized that different licensing models at different points in the evolution of the telecom sector represent a series of policy and regulatory bargains in the context of shifting economic, technical, and political circumstances and that in the case of the licensing model for wireless in local loop (WLL), one "trade-off is the use of spectrum for free, in exchange for a stronger set of performance requirements for firms", while a second trade-off "has been to consider some of the concerns of the holders of wireless licenses, and to rework some of the agreements governing interconnection and division of revenues between wireline and wireless companies" (McDowell and Lee 2003: 379). While the permission granted to limited mobility no doubt was indicative of policy and regulatory bargaining, the crucial component in the way incentives came to be designed in this case appears to be the political environment rather than the application of any technical or economic expertise on the part of the regulator. The decisions to give spectrum 'for free' cannot be supported on technical and economic grounds as late as in 2001;[10] it placed in relief the leadership role of the executive branch and the acquiescence of the independent regulator in not penalizing breaches of contract and perhaps even in encouraging the licensees' propensity to commit contract infractions.

The usage charge regime based on a share of revenue without insisting upon an upfront fee reflecting the scarcity value of spectrum thus came to be firmly established in the telecom sector around the turn of the millennium. The fourth CMSP licenses were tendered for on the same basis, as was the allocation on 'continuous basis' of FSP licenses after January 25, 2001, on the payment of an administratively determined entry fee.[11] There is a tendency to consider the fourth CMSP tenders as having yielded an upfront price for the spectrum that was implicitly bundled with the license, as for example in the assertion that "the fourth operator license was issued in 2001 using a three-stage auction procedure and a start-up spectrum of 2x4.4 MHz in 1710–1785 MHz paired with 1805–1880 MHz band was given to the winning bidder" (Sridhar 2012: 112). This standpoint is consistent with the discourse that holds in the context of the first round of CMSP and FSP tenders that "India was among the early adopters of spectrum auctions" (Jain 2001: 672). There are two problems with such a view: (1) first, the assumption that the bids were intended to invest property rights in spectrum with the winning bidder was nowhere specified; as noted, this was not even remotely the intention; and (2) the promise of some 'start-up' spectrum was a result of being locked in to the legacy of the first two cellular licenses that were given away on a trial-and-learning-by-error process that evolved from the metro tenders to the circle tenders; it was at best a way of ensuring a level playing field between incumbents and later entrants. A more perceptive analysis will recognize that in 2001 – once again – what was being offered through tender was the ITA section 4 privilege to establish, maintain, and work telegraphs, albeit with more clarity on the terms and conditions of the contract occasioned by the learning from prior experience. The tendered entry fee for CMSPs did not include any upfront payment for spectrum in the same way that the administratively determined entry fee for FSPs did not. While it is tempting to blame the regulator for "hesitance in choosing auctions, while agreeing in principle regarding the appropriateness of the method" (ibid: 676), such criticism is misdirected since TRAI did not even construe the CMSP tenders as investing property rights in spectrum in the potential licensees, quite consistent with its position in the case of FSP licensees. Without the analytical confusion that views the CMSP tenders as spectrum auctions, it is no longer true that the "two services that both require spectrum (incidentally in adjacent bands) [were] treated differently" (ibid). The misconception that the fourth CMSP entry fee reflected the scarcity and/or market value of spectrum is widely held and has been responsible for causing mischief and bringing much grief to the sector in later years. The market did not yield any upfront spectrum price in that tender; nor could it have, for spectrum availability was being assured by the government to the winning bidder,[12] who only needed to pay spectrum usage charges (SUC) in lieu of the assured supply and could, market conditions permitting, pass on this variable cost to the consumers.[13] The view that the bids made by service providers who won the 2001 licenses were a proxy for their valuation of the spectrum that came bundled with it is also responsible for the misconception that, despite the apparent success of the auctions, the government "parallely pursued the practice of assigning licences (including the spectrum that came bundled with it) on receiving applications from the providers on a first come first served (FCFS) basis" (Kumar 2011: 122).[14] Government pursued the practice of FCFS precisely because the spectrum valuation had not been undertaken and because only usage charges were being recovered from service providers upon granting usage rights at executive discretion. Interestingly, it conceded as much in 2006 to the Parliamentary Standing Committee by stating in its action taken report that "at present, there is no separate entry fee for spectrum or one time spectrum charge" and that "only spectrum usage charge is levied" (Standing Committee on Information Technology [SCIT] 06–07b: 25). To this extent, the executive branch was internally consistent in its approach, by default if not by design. The larger failure was on behalf of TRAI, which did not focus enough attention on the issues involved in discovering the scarcity value of spectrum even after it was mandated to do so by the amendment to the TRAI Act in January 2000 and instead continued to endorse the employment of an administratively determined variable cost mechanism to price this input resource.

The reluctance to move towards a market-based determination of spectrum value and price was no doubt a result of legacy issues. Nevertheless, TRAI lost an opportunity to reset priorities when it was called upon post-NTP 1999 to advise DoT on the introduction of competition. It failed to seize the opportunity to redefine the access services market, and it failed – perhaps as a consequence of its compartmentalized approach – to price spectrum appropriately. The 2 per cent base charge for spectrum had no discernible nexus with the scarcity value of spectrum. Spectrum was not only allowed to be administratively allocated, but also administratively priced, and this obviously had effects not only on QoS but also on the expectations of the service providers and their propensity to push group interests. The missed opportunity and TRAI's continuing regulatory dilemma were highlighted when DoT sought TRAI's views in December 2002 seeking recommendations on whether more CMSP licenses could be awarded. In its response on February 20, 2003, TRAI observed that in the context of the competitive scenario of four CMSPs and at least two FSPs in each service area, existing cellular operators in India had much less spectrum allocation in comparison to their counterparts in other countries and that "this imposes a needless constraint on them in respect to both QOS/optimal network engineering" (DoT 2011a: A215). However, TRAI still hedged its response by expressing its opinion that "induction of additional mobile service providers in various service areas can be considered if there is adequate availability of spectrum for the existing service providers as well as for the new players, if permitted" (ibid: A216), without grasping the bull by its horns and suggesting the method of allocation if indeed new CMSPs were to be inducted. DoT inferred from this recommendation that "it is TRAI's view that additional operators for CMTS should not be inducted at this stage" and that it was TRAI's view that "first, adequate spectrum should be ensured to the existing operators" (ibid: A220). Consequentially, DoT was also free to infer TRAI's endorsement of the SUC regime that was in place.

It has been noted that the introduction of unified access service (UAS) licensing in November 2003 that allowed FSPs to provide cellular access services with entry fee based on that paid by the fourth CMSP "theoretically allowed any number of mobile licenses to be provided and implicitly de-linked spectrum allocation from licensing" because even though "firms were awarded licence after paying the required entry fee, they were given start-up spectrum only as and when available" (Sridhar 2012: 113). This perception suffers from the same misconception that construed a linkage between license and spectrum in the tenders for the fourth CMSP. It also ignores the origins of the FCFS procedure for the issuance of spectrum license for FSPs seeking spectrum on obtaining operating licenses that were available on a 'continuous basis'. The UAS licensing regime, itself the culmination of the migration of FSPs into the cellular access market, had little focus on spectrum management issues; all the UAS regime did was import FSP-related processes such as continuous issuance of operating licenses and FCFS for spectrum licenses into the cellular access service licensing regime (now called the UAS regime). Spectrum and license were not coupled together in any meaningful relationship before UAS licensing, and the new regime did not alter the relationship in any significant manner. Supply of spectrum was assured by DoT subject to exercise of discretion in the earlier regime, and DoT continued to exercise that discretion by way of laying down SLC and additional SUC for allocating additional blocks of spectrum to existing operators as and when DoT said it became available. By basing the entry fee for UAS license in the difference between the entry fees paid by FSPs and the fourth CMSP in a service area, DoT was not identifying the value of spectrum, given that TRAI had very clearly recommended that "service providers migrating to the Unified Access Licensing Regime will continue to provide wireless services in the already allocated/contracted spectrum and no additional spectrum would be allocated only because of migration" (TRAI 2003c: 28). All the migration to UAS licensing did was unify the access market and make it predominantly cellular-access led, and any other interpretation that attributes a spectrum aspect to the process is not supported by the evidence.

Notes

1 BSNL started its business operations with effect from October 1, 2000.
2 The first private sector NLD operator, however, entered the market only in November 2001 when Bharti Telesonic signed the license agreement (Uppal et al. 2006: 23). The international long distance service was a

monopoly of the government-owned Videsh Sanchar Nigam Limited (VSNL), which was privatized in February 2002.

3 Bharti offered steep discounts on long distance mobile-to-mobile calls upon entering the NLD market, and BSNL had to respond to this threat by making matching cuts.

4 BSNL's maintained a steady market share in both cellular and fixed access services over the second phase. It had a market share of 17.34 per cent with 4.93 million cellular mobile subscribers out of a total of 28.44 million as of December 31, 2003, while this improved marginally to 19 per cent (32.72 million out of a total 172.33 million) as of December 31, 2007; in fixed services, BSNL's held 84.8 per cent of the market with 35.71 million of the 42.09 million subscribers in December 2003, which declined marginally to 80.79 per cent (31.71 million out of 39.25 million) by December 2007 (TRAI 2004a, 2008b). MTNL, the public sector provider in Delhi and Bombay, also showed similar trends (cellular: 1.46/1.71 per cent; fixed: 10.60/9.16 per cent).

5 The Delhi High Court has held that the IUC Regulation "is in the nature of a direction to service providers under s 11, rather than a true regulation made under s 36" (Raghavan 2007: 195).

6 TRAI observed that its earlier estimates of the required quantum of USL as 5 per cent of AGR in its recommendations pertaining to FSPs, CMSPs, NLD operators, and so forth would be retained because "the facts and figures have not undergone any major change" in the interim (TRAI 2001b: 28).

7 Shyamal Ghosh, who retired as secretary, DoT on May 31, 2002, was appointed as the first USO fund administrator on June 1, 2002.

8 Part-time TRAI member Rakesh Mohan dissented, arguing that the fund should be used only to fund universal access under stream I since obtaining disaggregated costing and revenue information SDCA (short distance charging area)-wise from each operator would be inappropriate and cumbersome.

9 Of the total of about INR 13.47 billion disbursed from the fund between 2002–3 and 2010–11, BSNL received about INR 11.64 billion, or 86 per cent (usof.gov.in/usof-cms/usof_subsidy_previous.html, accessed August 30, 2013), even though the scope of funding was later progressively increased to cover schemes other than streams I and II.

10 New Zealand conducted the first auction of radio spectrum as early as 1990, and the US Congress authorized spectrum auctions by 1993 (Prasad and Sridhar 2014: 89–90).

11 After the bidding in 2001, 17 CMSP licenses were issued, and 22 FSP licenses were issued between 2001 and 2003 on continuous basis (DoT 2011a).

12 Recall that the cellular guidelines of January 5, 2001, stated that additional bandwidth would be allotted subject to availability and justification and "shall attract additional license fee as revenue share" that would be reviewed by the Wireless Planning and Coordination (WPC) wing of DoT from time to time (DoT 2011a: A201). Not only the so-called start-up spectrum but supply of further tranches also was thus assured by DoT at its discretion. This stipulation once again indicated that the SUC regime was not intended to elicit the market price of spectrum.

13 An upfront spectrum charge on the other hand would have been a fixed cost or sunk cost depending upon specific features such as "the number of operators, the contestability of the industry, the degree of fragmentation of spectrum, and the nature of demand" (Prasad 2011: 25).

14 Kumar does recognize that FCFS was conditional upon the payment of the one-time entry fee determined by the 2001 auction; however, he does not delve deeper into the contradictions in the unified access service (UAS) licensing regime that we have pointed out.

6

THE COMPETITIVE SECOND PHASE (2004–7)

The cabinet decision of October 31, 2003, and the migration to the unified access service (UAS) licensing regime with many inherent contradictions still in place were the defining events of the period leading up to the intensely competitive phase of mobile telephony growth witnessed in the second phase. The cabinet decision was important not only for muting the contentions among different service providers and initiating the new licensing regime, but also for widening the ambit of policymaking in the telecom sector. It did this by mandating greater coordination among different departments of government with regard to spectrum management. The cabinet approved the group of ministers (GoM)'s recommendations that (1) adequate spectrum be made available for the unimpeded growth of mobile services, "modalities for which will be jointly worked out by Wireless Planning & Coordination (WPC) Wing of Department of Telecom and Defence services" (Department of Telecommunications [DoT] 2011a: A269); (2) the Ministry of Finance (MoF) will provide budgetary support to the Ministry of Defence (MoD) for modernization of their existing equipment to facilitate release of required spectrum, and the actual fund requirement will be worked out between MoF, MoD, and DoT; (3) DoT and MoF "would discuss and finalise spectrum pricing formula which will include incentive for efficient use of spectrum as well as disincentive for sub-optimal usages" (ibid: A270); (4) the allotment of additional spectrum be transparent, fair, and equitable, avoiding monopolistic situations in spectrum allotment and usage; and (5) long-term spectrum requirements and time frames would be worked out by DoT.[1] The decision was thus the first intimation of deliberative thinking on the considerable scarcity value of spectrum. It is a different matter that some of the stipulations imposed by the cabinet decision became the subject matter of inter-departmental conflict during the later part of the second phase. It is also interesting that the cabinet decision did not care to specify any role for the Telecom Regulatory Authority of India (TRAI) in the matter and left it to the discretion of DoT to involve the regulator in the decision-making process as it deemed fit.

The core of the executive decision to introduce the UAS licensing regime, however, was received with considerable enthusiasm, especially by those service providers who now possessed licenses to provide cellular access services with a pan-India footprint, along with national long distance (NLD) and international long distance (ILD) licenses that allowed them to access wholesale markets as well as provide direct interconnectivity among group operations.[2] Access service licenses could now be obtained based on intending operators' capability and market perception in each of the service areas. Fixed service providers (FSPs) were quick to complete migration to the UAS licensing regime and started offering cellular access services soon after the guidelines were announced on November 11, 2003. Table 6.1 indicates the competitive snapshot for access services in each service area on March 31, 2004, and in January 2005. It is apparent that competition had arrived, and arrived in a big way, in most large service areas. Along with competition came lower retail prices and a step up in consumer demand for access services. The competitive pressure resulted in a spurt in the demand for spectrum, particularly in larger cities and urbanized states. We noted earlier that subscriber linked criteria (SLC) were devised by DoT to base assignment of additional spectrum to licensees; the SLC were naturally imprecise and had the effect of misaligning incentives for efficient spectrum use.

DoT acted completely on its own in setting up the SUC regime and the spectrum allocation procedure as of 2001, and TRAI had little if any role to play. The 'case by case' approach favoured by DoT is consistent with executive authority everywhere; the political nature of decision-making on a key economic issue also saw full play of group interests, as expected under the plea of shoring up investor confidence. What is of particular interest in this case is the existence of an independent economic regulator which was quite vigorous in exercising jurisdiction in some matters such as retail tariffs and supporting the government's initiatives in transforming the licensing framework but showed no enthusiasm in coming to grips with as important an issue as spectrum pricing. TRAI had no role in the way the SUC regime was initiated and continued, and it did not show any regulatory concern for developments on this front until it made its *suo motu* recommendations on unified licensing on October 27, 2003. The task of efficiently managing available spectrum was borne entirely by DoT in the interim, with no regulatory input from TRAI, despite the statutory

Table 6.1 Access service licensees in the second phase

	Service area	*Number/Names of licensees providing access services (fixed, mobile, wireless in local loop [WLL] (F) and WLL (M))*		*Remarks*
		At the end of first phase	*As of January 2005*	
Metros				
1	Chennai	6/ Bharat Sanchar Nigam Limited (BSNL), Reliance*, Bharti, Aircel, Hutch-Essar, Tata*	6/ As before	
2	Delhi	6/ Mahanagar Telephone Nigam Limited (MTNL), Reliance*, Bharti, Tata*, Hutch-Essar, Idea	6/ As before	
3	Kolkata	5/ BSNL, Reliance*@, Bharti, Hutch	6/ Earlier list + Tata (2004)	
4	Mumbai	6/ MTNL, Reliance*, Bharti, Hutch-Max, BPL, Tata*	6/ As before	
Category A				
5	Andhra Pradesh	6/ BSNL, Reliance*, Tata*, Idea, Bharti, Hutch-Essar	6/ As before	
6	Gujarat	6/ BSNL, Reliance*, Tata*, Fascel-Hutch, Idea, Bharti	6/ As before	
7	Karnataka	6/ BSNL, Reliance*, Bharti, Tata*, Spice, Hutch-Essar	6/ As before	
8	Maharashtra	6/ BSNL, Reliance*, Tata*, BPL, Idea, Bharti	6/ As before	Included Goa
9	Tamil Nadu	6/ BSNL, Reliance*, Bharti, Tata*, BPL, Aircel	6/ As before	

Category B				
10	Haryana	5/ BSNL, Reliance*, Bharti, Escotel, Aircel	6/ Earlier list + Tata (2004)	
11	Kerala	5/ BSNL, Reliance*, Escotel, BPL, Bharti	6/ Earlier list + Tata (2004)	
12	Madhya Pradesh	6/ BSNL, Reliance*@, Bharti*@, Idea	7/ Earlier list + Tata (2004)	Included Chhattisgarh
13	Punjab	6/ BSNL, Reliance*, HFCL*, Spice, Bharti, Hutch-Essar	7/ Earlier list + Tata (2004)	
14	Rajasthan	6/ BSNL, Reliance*, Shyam*, Aircel, Hexacom, Idea	7/ Earlier list + Tata (2004)	
15	Uttar Pradesh (East)	4/ BSNL, Reliance*, Aircel, Escotel	6/ Earlier list + Tata, Bharti (2004)	
16	Uttar Pradesh (West)	4/ BSNL, Reliance*, Escotel, Bharti	6/ Earlier list + Tata, Hutch-South (2004)	Included Uttaranchal
17	West Bengal	3/ BSNL, Reliance*@	7/ Earlier list + Tata, Bharti, Hutch, Dishnet (2004)	Included A & N, Sikkim
Category C				
18	Assam	2/ BSNL, Reliance*		
19	Bihar	3/ BSNL, Reliance*@	6/ Earlier list + Tata, Bharti, Dishnet (2004)	Included Jharkhand
20	Himachal Pradesh	5/ BSNL, Reliance*@, Bharti, Idea	7/ Earlier list + Tata, Dishnet (2004)	
21	Jammu and Kashmir	1/ BSNL	4/ Earlier list + Bharti, Reliance, Dishnet (2004)	
22	North East	3/ BSNL, Reliance, Bharti	4/ Earlier list + Dishnet (2004)	Included six states
23	Orissa	3/ BSNL, Reliance*@	6/ Earlier list + Tata, Bharti, Dishnet (2004)	

* FSP licensees who later opted to migrate to UAS license; @ Company had both FSP and CMSP licenses converted into UAS licenses
Source: Compiled from TRAI 2004c: 42, 53–59, 2005a, and www.dot.gov.in.

provision in section 11 (1) (a) (viii). Even when it gave its attention to spectrum-related issues in 2003, TRAI was content to endorse the DoT's regime. It specifically recommended on November 19, 2003, that "there shall be no change in the spectrum allocation procedure as part of the migration process" to the UAS regime, that if new entrants are allowed by the licensor, the existing procedure would apply to them as well, and that "spectrum under the new unified license for offering mobile services may be allocated in the interim period on the technology used for offering these services" (DoT 2011a: A711–712). This instance of accommodation on the part of the regulator served to legitimize the executive branch's decisions on spectrum management.

TRAI's unfinished agenda

The story of the defining institutional interplay in the second phase of regulatory governance begins with TRAI's unfinished agenda of putting in place the unified license (UL), of which the UAS licensing regime was only the first step. The regulator initiated consultations on the second stage to follow up on the earlier recommendations on UL/UAS given on October 27, 2003, by issuing papers on November 15, 2003, and March 13, 2004. In its recommendations given on January 13, 2005, TRAI indicated that the "key objective of the Unified Licensing Regime is to encourage free growth of new applications and services leveraging on the technological developments in the Information and Communication Technology (ICT) area" while simplification of licensing procedure, ensuring flexibility and efficiency in resource utilization, encouraging efficient small operators in covering under-developed niche areas, and "ensur[ing] easy entry, level playing field and 'no-worse off' situation for existing operators" were listed as the other main objectives (TRAI 2005a: 1). Towards achieving these objectives, TRAI recommended that four categories of license may be introduced: (1) a unified license covering all public networks including switched networks irrespective of media and technology capable of offering voice and/or non-voice services including internet telephony, cable TV, direct-to-home, and television and radio broadcasting; (2) a class license covering services (including satellite services) that do not have both-way connectivity with the public network, such as radio paging and public mobile radio trunking systems; (3) licensing through authorization for providing passive infrastructure and bandwidth services; and (4) a standalone broadcasting and cable TV license. The first three categories were envisaged as being hierarchical, with unified licensing at the top. The other recommendations included: (1) the stipulation that the service areas remain the same as in the UAS regime, with operators having the option to obtain a national license; (2) rollout obligations similar to the UAS licensees; (3) a reduction in license fee to a uniform 6 per cent of a redefined adjusted gross revenue (AGR) that would include only revenue accruing from telecom services; and (4) a UL registration charge consisting of two parts – (a) a component for providing long distance services of INR 107 crore based on entry fee paid by NLD and ILD licensees and (b) a component based on a formula using the entry fee paid by new FSPs who entered in or after 2001 in each service area of UL access operations. TRAI also recommended that the UAS licensing regime would continue for a period of two years after the UL regime was put in place and that, on expiry of that period, all new licenses would be issued under the new regime.

The UL recommendations of January 2005 did not cover spectrum issues in detail since TRAI had separately initiated a consultation process based on DoT reference to it on November 17, 2003, seeking its views on efficient utilization of spectrum, spectrum pricing, and spectrum allocation procedure (DoT 2011a: 710). Nevertheless, it articulated the view in connection with spectrum pricing that "in the existing policy, spectrum charges have two components – (i) one time spectrum charges which are paid as part of one time entry fee by the service providers and (ii) annual spectrum charges which are paid in the form of percentage of AGR" (TRAI 2005a: 36). This assertion by TRAI ignored the history of access service privatization and was responsible for much later confusion in the matter of spectrum pricing when it was carried into the spectrum-related recommendations that followed in May 2005. As far as the introduction of the UL regime was concerned, TRAI recommended the continuance of existing spectrum pricing and allocation procedures until the government finalized new guidelines.

TRAI's UL recommendations of January 2005 were caught up in inter-departmental wrangling over the regulation of broadcasting. These recommendations were intrinsically linked to the fate of the Communications Convergence Bill that was already the subject matter of consultation between DoT and the Ministry of Information and Broadcasting (MIB) pursuant to the cabinet decision of October 31, 2003. The UL recommendations were referred by DoT to MIB for their views, and MIB responded with concerns regarding unification of content and carriage regulation and other issues.[3] A long and tortuous inter-departmental back-and-forth followed until the government decided not to accept the TRAI recommendations of January 2005 on July 10, 2007. While the recommendations did not receive acceptance, many of the principles advanced by TRAI in support of its views and the categories introduced for consideration survived in regulatory discourse and played an important role in subsequent developments in the governance of the sector.

The most important recommendations made by TRAI May 2005 for the purpose of this narrative are those pertaining to spectrum pricing. In addition to the formulation advanced in the UL recommendations of January 2005 that spectrum charges in the existing licensing framework consisted of a one-time spectrum charge (included in the entry fee) and annual SUC, the spectrum pricing recommendations also clarified that the "one-time entry fee/Registration charges recommended in Unified Licensing [did] not include spectrum charge" given that those recommendations envisaged delinking of license and spectrum (ibid: 76). TRAI then went on to recommend that "in the existing framework the spectrum charges should continue to have two components: one time spectrum charge and annual spectrum charge" (ibid). TRAI also observed that "the existing service providers have already been allocated spectrum beyond 2 x 6.2 MHz as specified in the license agreements without charging any extra one-time spectrum charges" (ibid: 77). The contradiction in the framework that appeared to TRAI to have charged a one-time spectrum charge for start-up spectrum but not for incremental spectrum escaped the regulator's attention (there was in fact no contradiction, as already noted, since the entry fee did not include any upfront spectrum charge). This misapprehension of facts directly led to the convoluted logic that followed in the regulator's recommendations. First, TRAI recommended that "there will be no one time spectrum charges for allocation of IMT-2000 [International Mobile Telecommunications 2000] spectrum to the existing service providers" since 3G (third generation) spectrum allocation "should be viewed as extension of 2G [second generation] spectrum allocations" and since revenue generation through sale of spectrum was not a key objective in its perception (ibid). Second, TRAI acknowledged that "in UASL, the one time spectrum charges and entry fee for license have not been separated" and concluded that this implied that the entry fee included a one-time spectrum charge also (ibid: 78). To get around this mix-up, TRAI then recommended that after the UL regime was implemented by the government, "the one time spectrum charge would be equal to UASL entry fee in that service area minus the component of registration charge based on the entry fee paid by the new BSO [basic service operator] (entered in/after 2001) specified by TRAI in its recommendations on Unified licensing regime dated 13th January 2005" (ibid), apparently to avoid the double counting of entry fee in the case of new entrants.[4] As far as SUC was concerned, TRAI noted the escalating charges in operation for Group Speciale Mobile (GSM) licensees varying between 2 and 6 per cent of AGR and the 2 per cent flat charge for spectrum up to 2 × 5 MHz for Code Division Multiple Access (CDMA) licensees and, after an evaluation of the pros and cons of alternatives, recommended that the "existing method of annual spectrum charge in terms of percentage of revenue share should continue" and that the highest SUC should be brought down to 4 per cent of AGR subject to periodic review (ibid: 81–82). On the quantum of SUC for the impending IMT-2000 spectrum assignment, TRAI declared that it was "not in favour of putting any extra annual spectrum charge", but in view of the potential for hoarding, it proposed that a "minimum guarantee of spectrum charge be taken for the period till the service provider rolls out IMT-2000 services" (ibid: 82). On the procedure to be followed for allocation in case demand outstripped supply, TRAI recommended that unless 2 × 5 MHz in the 2 GHz IMT-2000 band was sufficient to meet the demand from each existing operator, no spectrum allotment should be made. If the government did not favour this approach, only a "bidding option" for the minimum guarantee component should be followed (ibid: 84). Where supply equals or exceeds demand, the additional per MHz charge (minimum guarantee) for IMT-2000 spectrum "should be on the basis of highest charge per MHz per annum paid by any operators in different service areas" (ibid: 86).

This complex spectrum pricing framework was problematic in more ways than one. The most obvious problem was the way it assumed that the entry fee for existing licenses also had a one-time spectrum charge component, as noted earlier. The second problem was its endorsement of a flawed policy of assignment of incremental spectrum that was based on executive discretion approximated by subscriber numbers; the SLC not only served to misalign incentives, as service providers tended to inflate subscriber numbers in order to become eligible for incremental spectrum, but also vested considerable leverage in DoT's hands in making spectrum assignment decisions. Finally, the framework envisaged that spectrum for 3G would also be given out to service providers in the same way the incremental spectrum was being assigned – that is, based on executive discretion (in this case approximated by rollout obligations). In the face of international experience,[5] TRAI decided that "it is not advisable at this stage that a separate 3G-spectrum policy should be brought out" on the strength of a belief that 2G spectrum shortage could be mitigated with strategies that "shift some users from 2G bands to IMT-2000 band, thereby creating space for new and marginal users in the existing 2G bands" (ibid: 62). This was certainly an innovative regulatory strategy to overcome spectrum scarcity and engendered complacency in the establishment regarding the need to elicit market price for spectrum, independently of the debate regarding whether 3G spectrum was an extension of 2G spectrum. As seen later, the question of whether 2G and 3G spectrum could be differentiated on an economic basis itself became acutely important when 3G spectrum auctions actually took place in 2010. It can also be said that these recommendations made by TRAI in May 2005 did not have any immediate fallout for the telecom sector; it appears these recommendations were not even placed before the Telecom Commission for policy level consideration by the executive.

DoT issued a revised set of guidelines for UAL licenses on December 14, 2005, "in the light of enhancement of Foreign Direct Investment ceiling in [the] telecom sector" from 49 per cent to 74 per cent with effect from November 3, 2005 (DoT 2011a: 66).[6] The government's press note enhancing the foreign direct investment (FDI) limit required enforcement of certain conditions through amendments to license agreements (see ibid: A314–A318).[7] Apart from incorporating these conditions, the opportunity of issuing these guidelines was utilized by DoT to formalize two aspects of the licensing procedure that had no explicit regulatory logic or policy approval. First, the UAS guidelines of December 2005 stipulated at no. 11 that "licences shall be issued without any restriction on the number of entrants for provision of Unified Access Service in a Service Area" (ibid: A324). This guideline completed the importation of the 'continuous basis' of award of licenses from the FSP licensing framework into UAS licensing; it also took no reference to the availability or otherwise of spectrum that would obviously act as a constraint on issuing licenses "without any restriction on number of entrants". Second, to compound the error of opening up competition to an unspecified number of access service providers, the guidelines nowhere laid down the substantive procedure for making licensing decisions if the intending entrants exceeded the number of spectrum slots available.

The revised UAS guidelines of December 2005 thus sealed the policy and regulatory compact that had never taken into account the scarcity value of spectrum and had always reckoned that usage charges – initially paid on the basis of a formula but later based on a share of service providers' revenue – were sufficient payment for granting usage rights over the property of specified bands of electromagnetic spectrum. These guidelines also served to distance the licensing decision from the mandate of section 11 (1) (a) (i) of the TRAI Act that required TRAI to make recommendations on "the need and timing for introduction of new service providers"; with these guidelines that removed the restriction on the number of new service providers in place, DoT had no reason to seek TRAI's recommendations under that section. More ominously, the decision to place no restriction on the entry of competition itself was taken without even formally seeking the regulator's recommendations. In fact, given that TRAI had, on more than one occasion, stuck to its position that "induction of additional mobile service providers in various service areas can be considered if there is adequate availability of spectrum for the existing service providers as well as for the new players", the guidelines appear to have been issued in the face of this salutary – if ambiguous – exhortation from the regulator. It is also a measure of the executive branch's ingenuity that TRAI's consistently held position was interpreted under the December 2005 guidelines to mean that DoT had the exclusive jurisdiction to decide on a 'case by case' basis whether spectrum was available both for initial assignment as well as for meeting the incremental requirements of service providers. This interpretive ingenuity was to cost the sector heavily in the future both with regard to the number of licenses to be issued at a given point in time as well as in making spectrum pricing decisions. The stage was set for the later, contentious decisions in 2007–8 on the cut-off dates for UAS applications at the time these guidelines opened the application process on a continuous basis in 2005. Simultaneously, the guidelines also gave a fresh lease on life to the SUC regime, which did not factor in any upfront scarcity value for the spectrum being assigned and allocated to service providers. They also signalled that the second stage of unified licensing, which was to have overhauled the licensing and spectrum assignment and pricing framework, was put in cold storage. The complete isolation of TRAI from the spectrum-policy-making regime at this time was evident in the way a group of ministers was constituted on February 23, 2006, with terms of reference that impinged on an issue squarely within the ambit of economic regulation.

TRAI's recommendations of September 2006 on 3G spectrum were substantially different from those it made in May 2005. The regulator attributed the "new approach" to the "changes in the availability of spectrum and stakeholder inputs" since 2005 (TRAI 2006c: 9). It also discovered that spectrum is "a very scarce resource" and that a "serious crunch situation is on the horizon", making it imperative that the spectrum policy "must reflect this reality" (ibid: 25). The foremost change followed from the consequential observation that the limited "available spectrum" could be distributed among only a limited number of service providers and that "one of the most commonly used selection criteria of such eligible operators for allocation of spectrum could be on the basis of price to be paid by them" (ibid), thereby making the allocation decision the raison d'etre of the valuation of spectrum. What followed was the assertion that "3G cannot be perceived as an automatic extension of 2G and would need to be viewed as a kind of stand alone service for specialized needs and its allocation criteria has to be specified separately" (ibid). This assertion quite obviously overturned TRAI's May 2005 recommendation to treat 3G spectrum as an extension of 2G spectrum for allocation to existing service providers. More crucially, it set up an arbitrary divide between separate bands of the same electromagnetic spectrum for valuation purposes, as well as different allocation and pricing regimes for 2G and 3G spectrum only on the grounds of finding a suitable selection criteria in the face of a perceived scarcity of "available spectrum". By setting up different regimes for 2G and 3G spectrum, TRAI was ignoring the possibility that this could create "arbitrage possibilities . . ., administrative bottlenecks, and conceptual inconsistencies" (Prasad 2011: 25–26).[8] This half-way recommendation was also a show of the regulator's diffidence, engendered by the statutory mandate of section 11 (1) (a) (viii) that permitted it to make recommendations only on efficient management of *available spectrum*. The absence of statutory teeth was in sharp focus in TRAI's recommendations that identified the bands for rollout of 3G services: given that the regulator's 'request' to the defence users to consider vacating the 1850–1910 MHz paired with 1930–1990 MHz was not conceded and that the "Defence services do not foresee vacation of this band in the near future" (TRAI 2006c: 28), as well as the endemic shortages in the 800 MHz band it could do little about, TRAI could effectively make credible recommendations only with respect to the 2.1 GHz band in which 2 × 25 MHz band was available. TRAI was content to exhort that DoT should have "a time bound roadmap for making available additional and sufficient spectrum" (ibid: 3) and to suggest that a National Frequency Management Board be constituted to ensure proper spectrum management.

Fundamentally, however, the 3G spectrum recommendations are problematic because they distinguished between 2G and 3G spectrum on the rather specious basis that 3G is a "kind of" standalone service for specialized needs requiring allocation criteria to be specified separately. It built up this argument on the unsustainable grounds that selection of eligible operators in a situation where demand exceeded supply "could be on the basis of price to be paid by them". Scarcity was true of 2G spectrum as well, so the reason for the special allocation criteria for 3G was not clear from TRAI's recommendations. In fact, the recommendation to see 3G bands separately served to perpetuate the paradigm of the right to provide a particular service that had been the centrepiece of the licensing framework in India from the beginning and that – as noted earlier – evolved ever so slowly to bundle specific quantities of specified bands of electromagnetic spectrum with that right on the payment of usage charges in the nature of a variable cost to the right-holders. A rather more substantial basis for recommending the introduction of an upfront charge spectrum could have been a movement from the service provision right to the right to use, roughly "the right to transmit over a particular frequency band, over a particular geographic area" (Guasch and Spiller 1999: 229), something TRAI did not consider seriously[9] in giving its recommendations. All it could think up as a basis for introducing spectrum auctions was that 2G and 3G spectrum were two different creatures and demanded different allocation and pricing regimes. This reflected not only a paucity of imagination but also a debatable economic understanding of the spectrum resource.

With the benefit of hindsight, it must be said that TRAI's flawed reasoning has caused considerable mischief in the way the debate around spectrum as scarce natural resource has since been conducted via lobbying, audit, politics, and litigation civil and criminal. It may appear a tad unfair to criticize TRAI for what it did not say, but it is necessary to hold TRAI to account for lazy economic thinking with respect to differentiating between 2G and 3G spectrum. There is little in TRAI's recommendations to support the logic for auctioning 3G spectrum alone[10] and to continue with the prevalent regime for the 2G spectrum. In fact, the separation of the two was also perhaps spurious given that no fundamental change was contemplated in the property rights accruing to potential winners of the 3G auction; TRAI went too far in raising arbitrary boundaries between 2G and 3G spectrum in order to justify the auction mechanism for one and administrative pricing for the other. It would appear that the September 2006 recommendations were intended solely to counteract TRAI's previous recommendations of May 2005 that had viewed 3G spectrum as an extension of 2G spectrum and had sought to extend the SUC regime despite available international experience of 3G spectrum auctions. A more balanced appraisal of legacy issues would have occasioned a regulatory response to weightier economic questions such as those involving the vestment of a modicum of property rights (different from the existing service provision rights) in the spectrum band being put to auction. The lackadaisical consideration of *what* it was that was being auctioned – that is, whether a different regime of rights was being sought to be put in place for 3G spectrum and for what reasons – befuddled the recommendations and had important implications for the way they were later used and interpreted by the executive branch. TRAI recommended auction of 3G spectrum to elicit an upfront value (it called this a 'spectrum acquisition fee'), but also chose to retain the system of incremental SUC charges for the additional 3G spectrum that would be won by bidders. What was missing in the recommendations was serious consideration of introducing additional competition; the thread running through the recommendations was to find an allocation mechanism *for allocating 3G spectrum to existing operators*, and this was a significant departure from the many 3G auctions around the world TRAI quoted in support of its recommendations.[11] Thus, the September 2006 recommendations were not all that different from the May 2005 recommendations except for the assertion that 3G spectrum could not be seen as an extension of 2G spectrum and should be auctioned (among existing operators) instead of being administratively allocated. The auction recommendation was obviously intended to overcome the allocation problem; why this was limited to 3G spectrum alone while extant 2G shortages continued to be left in the hands of executive discretion cannot be explained by the text of TRAI's September 2006 recommendations.

Recommendations on capping access service licenses

DoT wrote to TRAI on April 13, 2007, seeking the regulator's recommendations on limiting the number of access providers in each service area as well as on reviewing license terms and conditions – both aspects the government was stated to be "contemplating" (TRAI 2007c: 145). The reference also conveyed the government's intention to review the terms and conditions in the access service licenses. The DoT's reference was significant not only for seeking to review its policy of issuing UAS licenses on a 'continuous basis' but also for articulating the government's preference for allowing license holders to provide services using different technologies under the same license. Dual-technology licensing (as this came to be referred to) was intended to ameliorate the disadvantages of the path-dependency of CDMA operators given the preponderance of GSM-based wireless subscribers in the market.[12] In seeking to limit the number of service providers, DoT was trying to come to terms with legacy issues that had seen the importation of the 'open queue' principle from FSP licensing that had not reckoned on the burgeoning value of the spectrum resource into wireless-led access service licensing as discussed earlier.

TRAI responded to DoT's reference in its recommendations of August 28, 2007 (TRAI 2007c).[13] These recommendations hold considerable importance in the regulatory narrative for a range of issues and deserve attention for setting the agenda that unfolded in the third stage. On the issue of determining a cap on the number of service providers in a service area, TRAI noted that the main issue was whether the optimum number of service providers was to be determined by policy or whether to "leave it for market forces" (ibid: 6). It also noted that as of July 2007 there were 180 licensed cellular and UAS licensees in the country but the high level of competition in licensed service areas had not affected the widening gap between urban and rural teledensity. Crucially, the regulator saw "availability of spectrum, its allocation criteria, pricing methodology and method of evaluating utilization of the spectrum as to its technical and economic efficiency" as central to a proper appreciation of stakeholder comments regarding the imposition of a cap since uncertainty over availability had led to polarization of viewpoints on the issue (ibid: 15).

TRAI sought to test the conflicting viewpoints arising from the consultation process against (1) NTP 1999, (2) growth objectives, (3) sector experience including financials of existing operators, (4) principles of fair competition and other vital economic criteria, and (5) the direction of technological developments and concluded that "it is not advisable to exogenously fix the number of access service providers in a market which is in a dynamic setting" (ibid: 26). In arriving at this conclusion, TRAI read the provisions *related to CMSPs* in NTP 1999 to note that "there is no intention of placing any artificial cap on the number of access service providers" and to assert that the clear, underlying policy theme was "to ensure optimality for existing operators so as to provide good quality service but at the same time it has not barred entry of new operators" (ibid: 17–18). Whatever the other virtues of not placing a cap on the number of access service providers, this interpretation of policy is once again guilty of brushing over legacy issues in licensing. TRAI had envisaged unified licensing as the second step in its considered recommendations of October 27, 2003, and this had been written into policy in the addendum issued to NTP 1999 on November 11, 2003; in this framework, additional competition would be introduced through unified license holders, with spectrum being delinked from license. TRAI's August 2007 interpretation of NTP 1999 as not barring the entry of new operators notwithstanding the non-implementation of the second step to unified licensing mandated by policy put the seal of approval on a flawed licensing framework that itself had emanated from a series of missteps suggestive of facilitating FSPs' foray into cellular access service without admitting to faulty policy design that categorized the access market rather arbitrarily in the first place. TRAI in fact noted that limiting the number of access providers would be construed as a retrograde step in view of its recommendations on unified licensing of January 13, 2005, that were meant to encourage "free growth of new applications and services leveraging on the technological developments in the information and communication technology sector" (ibid: 25). Strangely, however, the regulator did not appear to have taken note that those recommendations had been given short shrift by the executive and that there was no possibility of achieving objectives that were so tied in with convergence that they had effectively been jettisoned by the government by the time of August 2007 recommendations. TRAI's recommendation was borne of an imperfect understanding of both legacy and context. By recommending that "no cap be placed on the number of access service providers in any service area" (ibid: 27), TRAI perpetuated the confusion engendered by successive faulty readings of the policy mandate and compounded it by refusing to acknowledge the absence of progress on unified licensing. Implementation of the unified licensing framework would have decoupled license and spectrum and thereby afforded a more practical means of issuing licenses to potential entrants. TRAI, even as late as August 2007, failed to press for a logical conclusion to the process of unified licensing that it had recommended in 2003 and that more importantly had the imprimatur of the cabinet. Once again, TRAI was culpable for a grievous act of omission.

Instead of pressing DoT to honour the letter and spirit of the cabinet decision of October 31, 2003, that had been subsequently incorporated into policy by the addendum to NTP 1999 issued on November 11, 2003, TRAI went on a roving examination of the need to review spectrum management policies and to exhort the licensor to place information on spectrum allocation, pending requests, reasons for pendency, and so on in the public domain. It also recommended that "the likely availability of spectrum in different bands in all the circles, the time period by which it will be available for allocation, [and] the criteria to be adopted for setting up a [queue] system needs to be made public" (ibid: 27). While unexceptionable as a regulatory initiative, this recommendation saw TRAI go down the path of examining spectrum efficiency enhancement, allocation criteria, efficient pricing, and so forth, which finally led it to the familiar field of SLC and away from the possibility of envisaging the best solution to the problem of ensuring competitive service provisioning in an environment of spectrum shortage. In making its spectrum-related recommendations, TRAI was conscious that spectrum allocation followed grant of UAS license in the extant regime and that the licensee had no choice but to pay the entry fee for license in full. TRAI also observed that the entry fee determined in 2001 (for the fourth CMSP that was also used as entry fee for migration to UAS) "does not bear any relationship to present spurt in the telecom market" (ibid). On these grounds, it was recommended that "DOT should examine the issue early and specify appropriate license fee for UAS licensees who do not wish to utilize the spectrum" (ibid).

This was a curious formulation coming from the regulator, being neither the fish of UAS license nor the fowl of unified license. It was indeed a good red herring that misrepresented market reality and licensing practice and further confused the regulatory construal of the relationship between license and spectrum *in the UAS licensing regime*. While TRAI recognized the ambiguous relationship between license and spectrum at that point in time, it chose to put the ball in DoT's court; it wanted the licensor to "examine" the issue and "specify" license fees without making any actionable recommendations to unravel the knotty spectrum-license conundrum. This revealed regulatory reluctance to follow through on its earlier recommendation to move to a unified licensing regime. At best, this reluctance was a capitulation to political reality; a more critical view would see this as an abdication of regulatory responsibility since it devolved on TRAI to identify impediments to the orderly growth of the telecom sector and rectify them in time, especially since the unified licensing framework was a *suo motu* regulatory initiative undertaken by it in 2003 ostensibly to "[remove] barriers to facilitate growth in the sector" (TRAI 2003c: 18) such that a litigation-free environment would be created "because all service providers will be in a position to offer all types of services in all service areas depending upon the service provider's choice" (ibid: 31). The unified licensing structure remained unfinished since only the first stage UAS licensing had been introduced, but TRAI did not care to take its agenda forward in 2007 to decouple spectrum and license, adding yet another episode to the trope of chances TRAI missed in the course of its regulatory journey. That delinking had to await much turbulence and protracted litigation (exactly what the regulator had stated in 2003 would be avoided by recourse to the proposed unified licensing regime) before it was written into policy in NTP 2012.

TRAI next turned to the issue of spectrum pricing and usage charges, basing its examination on the premise that spectrum pricing should "provide incentive for efficient utilization and discourage creation of self-perpetuating shortages" (TRAI 2007c: 42). The first justification TRAI gave for not considering administrative or market-based spectrum valuation was on the basis of distinguishing between entry fee and SUC. In an oft quoted and sometimes contested paragraph in the August 2007 recommendations, the regulator laid out its rationale for maintaining the status quo with respect to spectrum pricing in the 800, 900, and 1800 MHz bands that were deployed by cellular mobile telephone service (CMTS) and UAS licensees in either of the two technologies. It is worth quoting in full:

> The allocation of spectrum is after the payment of entry fee and grant of license. The entry fee as it exists today is, in fact, a result of the price discovered through a markets based mechanism applicable for the grant of license to the 4th cellular operator. In today's dynamism and unprecedented growth of telecom sector, the entry fee determined then is also not the realistic price for obtaining a license. Perhaps, it needs to be reassessed through a market mechanism. On the other hand spectrum usage charge is in the form of a royalty which is linked to the revenue earned by the operators and to that extent it captures the economic value of the spectrum that is used. Some stakeholders have viewed the charges/ fee as a hybrid model of extracting economic rent for the acquisition and also meet the criterion of efficiency in the utilization of this scarce resource. [TRAI] in the context of 800, 900 and 1800 MHz is conscious of the legacy i.e., prevailing practice and the overriding consideration of level playing field. Though the dual charge in present form does not reflect the present value of spectrum it needed to be continued for treating already specified bands for 2G services i.e., 800, 900 and 1800 MHz. It is in this background that [TRAI] is not recommending the standard options pricing of spectrum, however, it has elsewhere in the recommendation made a strong case for adopting auction procedure in the allocation of all other spectrum bands except 800, 900 and 1800 MHz.
>
> (ibid: 46–47)

The 2007 recommendation's primary justification was the burden of legacy that placed differing quantities of administratively allocated and differentially priced spectrum in three different bands in the hands of incumbent service providers that made the determination of a cut-off date for implementing auction difficult. It is noteworthy that the solution to this problem could have been sought in following through on the unified licensing framework, which (1) would have delinked spectrum from license and (2) would have not only sketched a migration path for the existing licensees but also allowed new entrants to obtain licenses and expand the supply side. TRAI – for whatever reason – did not press for the cabinet-approved unified licensing regime to be implemented in full. This is a glaring omission given that even while recommending the auction of spectrum in the non-2G bands, TRAI *nowhere specified if and how such spectrum would be delinked from the license*. Nor could it have, since there was no license category that could have accommodated a new entrant looking to enter the access services market without seeking allotment of 2G spectrum. This regulatory reluctance was of a piece with TRAI's inclination to allocate additional spectrum – whether in the 2G bands by administrative fiat or in other bands by auction – to *existing operators alone*. Seen in conjunction with its recommendation to not cap the number of access service providers, this reluctance presents a bit of a contradiction, but, as we have noted on more than occasion, overcoming contradictions has not been among TRAI's regulatory strengths. All that can be stated in defence of TRAI in this issue is that it did recommend that DoT examine the matter and specify a fee for a "UAS license not wishing to utilize spectrum", which was a non-existent policy category.

A third set of recommendations TRAI made in August 2007 that evoked more controversy related to the combination of technologies under the same license. It must be acknowledged that TRAI noted that the UAS license (remember that the 17th guideline of November 11, 2003, mandated that all applications for new access service licenses shall be in this category) "[did] not contemplate more than one technology choice being made by the licensee" and that "by clearly acquiring license and technology choice at the outset and then linking both initial as well as additional spectrum allotments to the same, clearly rules out any scenario where a licensee can acquire spectrum for both technologies in a single license" (ibid: 109). Thus, it was TRAI's "firm view that crossover allotment of spectrum is not permissible under the present UAS license" (ibid). TRAI recognized that this was a result of legacy since "GSM and CDMA operators were given initial frequency allocation from separate frequency bands . . . the growth path for these two technologies are separate and at present based on subscriber numbers" and that "licensees using one technology should not be assigned additional spectrum meant for the other technology to avoid legal complications" (ibid: 110). After articulating its views on crossover with such clarity, TRAI resorted to the argument of maintaining a 'level playing field' instead of attacking the coupling of license and spectrum which was the nub of the problem. It acknowledged that most developed countries had delinked spectrum allocation from license and even that many countries had an "auction policy of spectrum rather than pre-determined imposition of technology" (ibid: 114). The regulator, however, refused to take the hint this experience offered by way of vesting 'use rights' in spectrum, but averred that "such treatments in mid-course of the growth path are riddled with implications for competition in the market" (ibid), thereby giving evidence of a business-friendly regulatory approach rather than a more appropriate market-friendly one. To ensure the level playing field was not disturbed, TRAI opted to impart flexibility to licensees to get out of their technological path-dependency, albeit at a cost. Accordingly, it recommended that "a licensee using one technology may be permitted on request usage of alternative technology and thus allocation of dual spectrum" and that "such a licensee must pay the same amount of fee which has been paid by existing licensees using the alternative technology or which would be paid by a new licensee going to use that technology" (ibid: 116). This fee turned out to be equal to the entry fee for UAS license, marking another instance in the Indian telecom services sector of change of contract terms midway; whether this can be termed contract renegotiation is unclear since it was at the initiative of the licensor that the regulator took up the examination of the matter and there is little in the text of the recommendations that reveals licensees' preference one way or the other. The second entry fee according to TRAI was not only legally tenable but "also pass[ed] the test of level playing field" (ibid). However, TRAI still left open the question of whether the fee was to be reckoned as an upfront spectrum fee or as a fee for expanding the scope of license. The spectrum-license confusion was to continue by virtue of this recommendation, and this was worse confounded by elaborate stipulations regarding the applicable SUC regime for the spectrum that would be allocated to the crossover licenses and the order of priority among all licensees for spectrum allocation. TRAI patted itself on the back for its approach, which it labelled "new, liberal, innovative and [which] recognizes the market related developments in the context of technology" (ibid: 119), without acknowledging that permitting dual technology also had a complicating effect on the licensing regime and left intractable the problem of inseparability of entry fee and upfront spectrum charge TRAI itself had identified in the first place.

TRAI made other recommendations in its August 28, 2007, response to DoT's reference – on mergers and acquisitions (M&A), regulating equity stake inter se licensed service providers, and rollout obligations – but these are of limited relevance to the present narrative. The jointness of license and spectrum continued to be the base of TRAI's regulatory approach even in 2007, nearly four years after the cabinet had approved a path to decoupling the two. Even as it gained greater clarity over the dimensions of the problem over time, TRAI continued to go with the legacy spirit, which did not place any upfront value on the spectrum resource. If the entry fee could not be apportioned between license and spectrum, if SUC was the only relevant direct cost for spectrum borne by licensees, and if there was merit in spectrum auctions in bands that had no legacy issues to grapple with, it was incumbent upon the regulator to seek to mitigate legacy issues in the 2G bands by means other than offering sops to existing licensees, especially since *such an alternative was available by virtue of the cabinet approval for unified licensing*. In contemplating the regulatory path not taken, we are once again left to ponder the possibility of a less disorderly growth of the sector in the years following if only TRAI had fully grasped the issue of separating spectrum and license.

DoT's reference to TRAI on matters it was "contemplating" was made in the background of a number of UAS license applications pending with it.[14] DoT decided on July 17, 2007, that processing of all pending UAS license applications would be withheld until TRAI's recommendations were received (DoT 2011a: 69). After receipt of the recommendations and their examination in DoT, the Telecom Commission considered them in a meeting held on October 10, 2007. Among the important decisions taken by the commission were (1) to approve the 'no capping' proposal, with the rider that "the policy for reservation of spectrum for expansion of network of operator (enhancement of capacity) before spectrum is allocated to new licensees . . . will not be referred to TRAI" (ibid: A1139); (2) not to approve the recommendation to examine the issue of UAS licensees who do not wish to utilize spectrum and specify appropriate license fee for this category; (3) to agree that SLC needed review and that operators should use more efficient methods and spectrum optimization techniques and that the Telecommunications Engineering Centre (TEC) should examine the recommendation to form a multi-disciplinary committee; (4) to approve levy of spectrum enhancement charge for acquisition of spectrum beyond 10 MHz for GSM and 5 MHz for CDMA; the charge was to be double that recommended by TRAI; (5) to take the view that the recommendation that all spectrum excluding that in the 2G bands should be auctioned in the future "may not be considered" since it was "beyond scope of present reference" (ibid: A1140); (6) to await TEC's recommendations on SLC before deciding upon TRAI's recommendation regarding revision of SUC; (7) to approve usage of alternative technology by CMTS and UAS licensees using GSM or CDMA subject to payment of a fee "at least equal to the same amount of entry fee which has been paid by existing licensees using the alternative technology or which would be paid by a new licensee . . . to use that technology" (ibid: A1143) and also subject to the condition that spectrum in the alternative technology would be allotted only if the eligibility for spectrum in the other technology continued to be maintained; otherwise the spectrum in the other technology would be surrendered; to waive the entry fee for use of alternative technology in the case of BSNL and MTNL; and (8) to approve that license fee for crossover licenses would be levied on the combined AGR even as a separate revenue record would be maintained for the purpose of spectrum allocation. The M&A and rollout recommendations were also broadly approved by the commission. The commission's views were approved by the minister of communications and information technology with some modifications on October 17, 2007.[15] The minister also directed that, in view of the approvals, "pending requests of existing UASL operators for use of dual/alternate wireless access technology should be considered and they should be asked to pay the required fees" and that "allocation of spectrum in alternate technology should be considered from the date of such requests to WPC subject to payment of required fees" (ibid: A1150).[16] A press note was issued on October 19, 2007, setting out the terms of the approval (ibid: A608–A609).

Recall that TRAI's recommendations were made public on August 28, 2007. The process of examination within DoT included the constitution of an internal committee on September 21, 2007, and the Telecom Commission's meeting on October 10, 2007, that finally resulted in the decisions taken on October 17, 2007. DoT had, at the stage of seeking TRAI's recommendations, already indicated it was "contemplating" review of its policy to issue UAS licenses that until then had been given on a "continuous basis". This sequence of events was perfect for catalyzing interest among potential entrants to start the application process for UAS licenses. It is somewhat a measure of the perceptions regarding TRAI's regulatory stature and its acceptability to DoT that new applications started "pouring in", and 167 applications from 12 companies for 22 service areas were stated to have been received as of September 24, 2007, in DoT (ibid: A568) in the face of TRAI's recommendation *not* to cap the number of access providers in a service area. Given that the regulator did not appear to have heeded the licensor's hint regarding placing a cap on licenses, it is analytically significant that market players reacted as they did by rushing to file applications for licenses, almost certainly anticipating DoT would find a way around TRAI's recommendation. The reaction is significant for reflecting the low institutional credibility of the independent economic regulator whose existence was always premised on the enhancement of credible commitment the independent regulatory authority (IRA) would provide to private capital. The fate of TRAI's recommendation was sealed in the marketplace even before the DoT decision on it; what followed later – apart from the allegations of criminality that are the subject matter of court cases and not quite relevant to the present study – was merely a manifestation of the state of play in the sector that never really depended upon the regulator to provide a roadmap for its orderly growth. The absence of statutory responsibility for licensing, the repeated failure to tackle the license-spectrum problem, and the rather chequered history of recommendations often made to suit the licensor's inclinations ensured that TRAI's latest recommendation not to cap the number of licenses – this time perhaps *not* in line with the licensor's impulse – did not evoke confidence among investors that they could wait to make their investment decision in the sector. The perception of the regulator's subordination to the licensor was one important reason for the market reaction – a flood of applications to DoT seeking licenses to operate unified access services. The other was of course the absence of practicability in the recommendation not to cap the number of access service providers; as already noted, TRAI did not make any meaningful recommendations on how spectrum would be made available to new entrants (once again perhaps a consequence of absence of statutory mandate) even if licenses were to be continuously issued. The availability of spectrum – while the details were not widely available – was no doubt another important parameter that facilitated the decision made by potential entrants to put in their license applications in a rush. TRAI's inability to craft a credible path to 'continuous licensing' that simultaneously gave assurance of a modicum of certainty regarding spectrum allocation was a contributory factor in the process. That a cabinet decision offering a way out of the mess was available to aid the regulator (but sadly not utilized) makes its inability particularly prone to reproach.

The rush to hypercompetition

Perceptions of institutional relationship apart, prospective licensees continued to file applications with the licensor. DoT did not sit idle in the face of the license applications pouring in. On September 24, 2007, it moved to impose a cut-off date for receiving UAS license applications. This move was aligned with the intention expressed in the reference to TRAI in April 2007 to put a limit on the number of access service providers as a means of solving the spectrum-shortage problem. It also contained within it the possibility of arbitraging among potential entrants by dividing them into those who had already applied and those who had not prior to that date. Without speculating on motivations it can be said that the move had huge potential to misalign incentives, and indeed it served to create an even bigger problem than it had set out to solve. The proposal was to "announce a cut-off for receipt of UASL applications such that no new applications will be received after cut-off date till further orders" (ibid: A568), as good an example as any for a dodgy exercise of executive discretion in the absence of regulatory clarity. The date suggested was October 10, 2007, on the ground it gave 15 days, "reasonable time to all who wish to submit new UASL applications so that the decision may not be challenged"; what was approved was to "close receiving applications on 1.10.2007, i.e., one month from the date of Trai's recommendations" (ibid). That this proposal entirely missed the point is easy to conclude after the event, but it appears that the contradiction in stating in the same breath that "it may be difficult to handle such large number of applications at any point of time" with regard to the 167 already received and to afford "reasonable time" to all who wished to submit applications appears to have entirely escaped DoT's attention at the time it was made and approved. The misalignment of incentives this represented was to prove gravely deleterious to the health of the sector.

The fallout of the rush to decide upon limiting the number of applications even while TRAI's recommendation for not capping the number of access service providers was accepted was obvious enough. The number of applications filed with DoT reached 575, with applications being made by 46 companies for licenses in 22 service areas, even as the government was "yet to decide on processing of these pending applications for grant of UAS licenses based on availability of spectrum and other issues" (ibid: A612). This fact was brought to the attention of the secretary of the Department of Legal Affairs of the central government by DoT in a letter dated October 26, 2007, seeking legal opinion to deal with what it referred to as an unprecedented situation whereby "the number of applications are very large and spectrum is limited and it may not be possible for the Government to provide LOI/Licence/Spectrum to all applicants at all if the existing procedure is followed" (ibid: A613). The letter expressed the worry that the "existing procedure of sequential processing will also lead to inordinate delays depriving the general public of the benefits which more competition will bring" and the "need to adopt a methodology which synchronize[s] with legal base in processing of pending UASL applications and allotment of spectrum to various categories of spectrum seekers" (ibid). It offered three alternatives: (1) to process the applications on a first come first served (FCFS) basis in the order of their receipt, issue letters of intent to eligible applicants depending upon the availability of spectrum in different service areas, stipulate deadlines for compliance and payment of fee, and so on, and then place the complying applicants in a queue based on their application date for issue of license (letters of intent, LoIs, of those who do not fulfil conditions would be terminated); other applications would be considered for issue of LoI if spectrum was available; (2) to issue LoIs to all those who had applied by September 25, 2007 (date on which the cut-off date was publicly announced) with seniority for license and spectrum to be determined either on the basis of date of application or the date and time of fulfilment of LoI conditions; or (3) to issue LoIs to all eligible applicants received up to October 1, 2007, simultaneously, stipulating that spectrum allocation would be subject to availability and not guaranteed, such that "LOI holders are supposed to pay the entry fee if their business case permits them to wait for spectrum allocation subject to availability and Initial roll out using wireline technology" (ibid: A615). The letter also indicated that a fourth alternative could be "any better approach which may be legally tenable and sustainable for issuance of new licences" (ibid). Additionally, the reference also sought to obtain legal opinion on whether applications for dual technology received after October 18, 2007, ought to be placed prior to existing applications or after processing all 575 applications.

The ingenuity or lack of it in the suggested alternatives is not pertinent to the present analysis. Neither is the subsequent course of events that led to DoT's withdrawal of the reference for legal opinion on the matter.[17] What is institutionally relevant is the substance of the reference and whether it was not amenable to a regulatory resolution rather than a legal one. Even a cursory examination of the alternatives referred for legal opinion indicates the techno-economic nature of the problem at hand that was perfectly amenable to resolution within the regulatory and statutory framework. It is a measure of the

IRA's institutional stature and credibility even more than the policy-maker's reluctance to work the statute to obtain sustainable regulatory resolution of the problem that is glaringly obvious from the way DoT sought to refer the licensing problem for legal opinion. It is also obvious that DoT's initial preference for capping the number of access service providers was perceived by potential entrants as the general policy orientation in spite of the regulator's recommendation to not place any restriction on competition, especially since TRAI had not indicated a regulatory path for spectrum allocation. The market bet on an impending embargo on issuance of licenses, and the rush of applications that followed complicated the already confused scenario. Under the circumstances, one reasonable course of action would have been for the licensor to have referred the matter back to TRAI for its recommendations under section 11 (1) (a) of the TRAI Act since the issues for consideration included the need and timing of introduction of new service providers, the terms and conditions of their license, and facilitation of competition and promotion of efficiency in the operation of telecom services, as well as the efficient management of available spectrum as covered by clauses (i), (ii), (iv), and (viii) of the statute. It will not suffice to argue that DoT was acting upon TRAI's recommendations on these issues that were received only three months earlier since the material conditions between the reference made in April 2007 and the time of the decisions regarding the processing of pending license applications had altered considerably by virtue of the rapid rise in the number of applications and the announcement of a cut-off date for their receipt that implied a repudiation in effect of the core TRAI recommendation not to place a cap on the number of service providers. While pending applications for crossover technology had been converted into letters of intent, more applications had been filed leading to a further complication in the reckoning of the order of priority for spectrum allocation. At the heart of the questions in the reference for legal opinion was the vexed license-spectrum linkage on which TRAI had offered no clarity in its previous recommendations. In fact, DoT had noticed this problem at the time of the Telecom Commission meeting on October 10, 2007. It had observed that DoT "had made a reference to TRAI to examine the issue of limiting the number of access service provider[s] basically due to the shortage of spectrum" and that "TRAI has not given any solution to the paucity of spectrum while recommending no capping on number of Access Services Licences" (ibid: A590). This was based on the report of DoT's internal committee that too had noticed this problem and had suggested that "the recommendation may be referred back to TRAI for policy for reservation of spectrum for expansion of network of operators (enhancement of capacity) before spectrum is allocated to new licensees" (ibid: A574–A575). This suggestion was ignored by the commission that observed that while "the issue of spectrum availability cannot be ignored while granting new licences", in view of "the central aim of NTP'99 to ensure rapid expansion of teledensity, the recommendation that no cap be placed on the number of access service providers in any service area is accepted" (ibid: A590–A591). This was a disingenuous recourse to policy stipulation in the face of the acute shortage of available spectrum that triggered the reference to TRAI in the first place. In the final analysis, the decision not to make a back reference to TRAI on this issue proved shortsighted and disastrous.

That said, it must also be remarked that DoT's approach was not notable for any greater clarity on the policy options to unravel the license-spectrum problem than that exhibited by TRAI in its August 2007 recommendations. It would appear the licensor's concern was not on evolving a policy response to issues of competition and efficiency but merely to find an expedient way around the shortage of spectrum without reference to the structural flaws in the licensing framework that had crept up over time as a result of ad hoc tinkering. As already noted, continuous licensing of wireless access licenses was also an expedient move introduced in 2003 (and further strengthened by the December 2005 guidelines) and was not informed by careful regulatory reflection or supported by explicit executive approval at the highest levels. It came into existence via the 17th guideline that received ex post regulatory concurrence without required consultation and serious application of mind and became a licensing fait accompli. Given this state of affairs, it would have been proper for DoT to take into account the existence of cabinet approval for steps to delink spectrum from license – that is, to move towards unified licensing after obtaining TRAI's recommendations in this regard. DoT could have done this either by making a reference back to TRAI in terms clearer than those suggested by its internal committee[18] or by moving on its own to effectuate the delinking if it thought TRAI's recommendations on the subject were received. The most obvious solution was a move to unified licensing without bundling *any* spectrum with the license and a concerted effort to draw up a plan to release additional quantities of spectrum for telecom services. No doubt this would have required consultation at the inter-departmental level and perhaps even consideration by the cabinet. More crucially, any suggestion to move to unified licensing would have had to take recourse to institutional memory that went beyond the letter of policy; recall that the addendum to NTP

1999 issued on November 11, 2003 (that followed from the cabinet decision of October 31, 2003) had merely introduced two categories of license (UAS and unified) without articulating their stage-wise nature that animated the regulator's recommendations in this regard. It is not clear whether such institutional memory was either missing or was not allowed to intervene in the dramatic rush of events. With the benefit once again of hindsight, it is again easy to argue that adopting such a course of action would have obviated the disorder that followed in the aftermath of the decision to stick with an arbitrarily assigned cut-off date and process all pending application until that date at one go. In another crucial omission, TRAI did not follow up on its chairperson's letter of October 19, 2007, that wished DoT to consider the August 2007 recommendations in their totality even though the press note issued by DoT on September 24, 2007, had clearly indicated a cut-off date for receipt of applications and even after the DoT's press release of October 19, 2007, carried no mention of the government's decision regarding its recommendation for a UAS license without spectrum. Once again, for TRAI to have taken a more forceful stand on unified licensing would have involved a deeper understanding of legacy, the nuances of the text of policy and policy changes, and the will to follow through on the regulatory path it had initiated *suo motu*. Perhaps such an exhumation of regulatory and policy history and the coordination of regulatory resources involved was asking too much of an institution that had seen only short bursts of proactive economic regulation since its inception.

The Ministry of Finance, however, expressed its reservations regarding the valuation of crossover licenses on November 22, 2007.[19] After referring to a presentation on spectrum policy made by the secretary of DoT to the cabinet secretary on November 20, 2007,[20] the finance secretary noted that a fee of INR 1,600 crore was determined pan-India for granting licenses to CDMA operators to crossover to GSM operations (approximately the quantum of entry fee paid pan-India for the fourth CMSP license and adopted as the entry fee for dual-technology licenses) and stated that "it is not clear how the rate of Rs 1600 crore, determined as far back as in 2001, has been applied for a licence given in 2007 without any indexation, let alone current valuation" (ibid: A619). He thought that MoF ought to have been consulted in view of the financial implications of the decision and sought a review of the matter, pending which it was suggested that "all further action to implement the above licences may please be stayed" (ibid). This letter elicited a response from the secretary of DoT on November 29, 2007, to the effect that "the entry fee was finalized for UAS regime in

2003 based on the decision of the Cabinet" and that "it was decided to keep the entry fee for the UAS licence the same as the entry fee of the fourth cellular operator" and further that TRAI, in its August 28, 2007, recommendations, "has not recommended any changes in entry fee/annual license fee and hence no changes were considered in the existing policy" (ibid: A620). The letter referred to the terms of the cabinet decision of October 31, 2003, that DoT may be authorized to finalize the details of the implementation of the UAS licensing regime with the approval of the minister of communications and information technology, "including the calculation of the entry fee depending on the date of payment based on the principles given by TRAI in its recommendations" in support of DoT's action.

DoT's response to MoF is problematic for the manner in which it drew attention to some parts of the cabinet approval of October 31, 2003, without reference to the category of unified license provided in NTP 1999. But this option had never been on the table once the raison d'être of UAS licensing as being the first stage of migration to unified licensing was buried in the text of the addendum to NTP 1999 issued on November 11, 2003. Neither DoT nor TRAI had the institutional wherewithal to activate the approval of the political executive in order to recover the licensing framework from the disfigurement it had suffered over time, even if that meant using the instrument of unified licensing that too was part of the distorting toolkit. That the *suo motu* resolution attempted in 2003 of the original problem of artificial categories of access licenses in the flawed policy document of 1999 had been left unfinished at the time acute spectrum shortage was identified as an impediment to orderly growth in 2007 is further evidence that the unification of access licensing was not part of an iterative improvement of the licensing regime but was motivated by other reasons, such as finding a quick fix to the clamour from FSPs to be rid of the straitjacket imposed by wireline technology. It also represents an imperfect institutional relationship between licensor and regulator, with neither party owning responsibility for the unified licensing decision; the regulator abandoned its recommendation once it became part of the executive decision-making process, and the licensor ignored the spirit of the recommendation once it was lost in the thicket of the verbose addendum to NTP 1999 that never expressly recognized that UAS licensing was to be the first stage in the journey to unified licensing. This is a manifestation of the structural imperfection in the TRAI Act that merely gives a recommendatory role to the regulator on issues absolutely vital to the project of signalling credible commitment, arising from the reluctance of the executive branch to shed any of its licensing powers in favour of the IRA. The non obstante clause at the head of section 11 (1) of the TRAI Act proved otiose in the light of the later, judicially determined division of that section into sub-sections (a) and (b) that rendered recommendatory functions ineffective in mitigating the effects of section 4 (1) of the Indian Telegraph Act (ITA) and its proviso that allowed the central government to grant telecom licenses "on such conditions and in consideration of such payments as it thinks fit". The conditions continued to be imposed at the licensor's instance in spite of the regulator's recommendations. The category of TRAI's recommendatory functions in the amended TRAI Act served to support the exercise of sovereign discretion based on exigency rather than act as the basis for the infusion of expertise into executive decision-making. The events of the later part of 2007 reinforce such a reading of the statute.

Notes

1 In addition to these five recommendations, the GoM also took note of the constitution (and progress) of an inter-departmental task force with representatives from WPC wing, DoT, MoF, MoD, and TRAI for finalizing the modalities for release of spectrum on October 6, 2003.

2 Other than BSNL, Bharti Airtel, Reliance Communications, and VSNL (Tata Communications) held NLD and ILD licenses since 2001 or 2002.

3 MIB's concerns persisted despite TRAI's clarification that the standalone broadcasting licenses would be issued by MIB and that MIB "would continue to be responsible for content regulation in broadcast services" (Raghavan 2007: 338). TRAI had already been vested with the regulation of broadcasting and cable TV services through a government notification dated January 9, 2004, that expanded the definition of telecommunication services under the proviso to section 2 (k) of the TRAI Act. Perhaps more relevant to the short shrift given to the Convergence Bill was the change in government after the general elections of May 2004 that saw a different coalition obtain parliamentary majority. The UL recommendations, too, with their heavy reliance on convergence, were sidelined in the process.

4 The imaginary double-counting is assumed on the basis of the TRAI chairperson's letter of November 14, 2003. Recall that this letter specified that the entry fee for new unified (UAS) licensee "would be the entry fee of the 4th cellular operator and in service areas where there is no 4th operator – the entry fee of the existing BSO fixed by the Government" (DoT 2011a: A305).

5 TRAI acknowledged that "almost all the countries in the world have treated allocation of 3G spectrum separately compared to the 2G spectrum through either auction or beauty contest or combination of both" (TRAI 2005c: 61). It did not go that route, arguing that there was spectrum scarcity in India and also that "this boundary of 2G or 3G spectrum will vanish" in the future (ibid).

6 While this is the received wisdom, the guidelines themselves do not refer to the FDI trigger. They begin by stating that they are being issued

"consequent to opening of Internet Telephony and merger of Chennai & Tamilnadu Service Area" (DoT 2011a: A319). However, they incorporate most of the FDI press note enhancing the FDI limit.

7 The conditions – mainly related to security – were further modified on April 19, 2007 (DoT 2011a: A347–A350).

8 TRAI did note that "it would be difficult to charge two different annual spectrum fees for 2G and 3G operations because it opens the possibility of arbitrage between two possibly indistinguishable revenue streams" (TRAI 2006c: 70), only to recommend thereafter that an additional (incremental) 1 per cent of AGR should be charged as SUC for 3G spectrum after a moratorium of one year.

9 TRAI did use the terminology of 'use rights' but loosely in its recommendations on broadband wireless access (BWA) spectrum. It noted that the WPC wing issued wireless telegraphy operational licenses, "required to use the spectrum on a yearly basis", and called these "spectrum use rights" (TRAI 2006c: 91–92). However, the WPC spectrum licenses were still service specific and did not allow licensees to 'use' the spectrum to provide services or deploy technologies as they thought fit.

10 Strictly speaking, auction was recommended only for spectrum in the 2.1 MHz band. Spectrum in the 450 MHz and 800 MHz bands was to be auctioned if the demand was more than the supply; otherwise the prices were linked to bid/reserve prices in the 2.1 GHz spectrum auction.

11 Even in the case of BWA spectrum, TRAI recommended that a "large part of BWA spectrum as identified should be allocated among UASLs, CMSPs, or Category A and B ISPs for circle level deployments", along with a reservation for smaller ISPs operating in areas with population less than one million (TRAI 2006c: 90–91). The pool of potential licensees was thus larger than only the UAS licensees who would bid for 3G spectrum. In other countries such as the UK (which was remarkably successful in generating high bids), there was a conscious attempt to design the auction process to incentivize the introduction of a new entrant. The importance of facilitating entry and designing a suitable auction process based on the particular circumstances in a national market have been emphasized in Klemperer (2002), referring to the less-than-successful subsequent auctions in other European countries. TRAI recommended auction of five slots of 2.1 MHz 3G spectrum in 2006 using simultaneous ascending auction; however, only two 3G spectrum slots were finally auctioned in 2010, yielding spectacular revenues to government. TRAI justified the priority being accorded to existing operators in its recommendations on the ground of seeking "a balance between competition, affordability, and accelerated growth of the sector" (TRAI 2006c: 48).

12 GSM subscribers, at 120.47 million, accounted for nearly 73 per cent of wireless subscribers, with CDMA subscribers (including WLL (F) subscribers) at 44.64 million making up 27 per cent as of March 31, 2007 (TRAI 2007e: 27–28).

13 There was a change in the minister at the helm of DoT between the time the reference was made in April 2007 and the time the recommendations were given in August 2007.

14 The number was over 50 by April 2007 (DoT 2011a: A362). There was some controversy surrounding why some applications were kept pending

while others had been cleared since it was noticed on May 18, 2007 that "certain applications submitted earlier [were] pending in some circles, and others submitted in other circles at a later date have been processed" (ibid: 367). The approval for the proposal to keep the applications pending until TRAI's recommendations were received was given on July 17, 2007. See also DoT 2011a: 100–105.

15 A notable modification was to direct TEC to submit its SLC report by October 31, 2007, and that, in the interim, existing operators becoming eligible, "as per TRAI criteria, may be released additional spectrum subject to maximum of 10 MHz per Circle" (DoT 2011a: A1150).

16 Interestingly, the pending requests for use of alternative technology were only from CDMA operators (for 20 service areas in the case of Reliance and one service area each from Shyam and HFCL) (DoT 2011a: A1151). This squares with our analysis regarding the attractiveness of crossover licenses primarily to CDMA technology operators in view of the larger uptake of GSM among subscribers.

17 For the record, the minister of law and justice opined on November 1, 2007, that given the importance of the case and the various options indicated, "it is necessary that the whole issue is first considered by an empowered Group of Ministers & in that process legal opinion of AG can be obtained" (DoT 2011a: A616). This advice was considered by DoT as "out of context" (ibid: A627), and approval was given on November 2, 2007, for processing all applications received until September 25, 2007 (the date of press notification) according to the existing policy and to leave the decision on the remaining applications for later. This decision process was marked by considerable internal contestation within DoT, first with the secretary of DoT drawing attention to NTP 1999 provisions regarding spectrum and asking for an examination of whether LoIs and UAS licenses "can be granted without assured availability of spectrum" (ibid: A630) since "only the cabinet can effect a change in the policy" (ibid: A624). This was followed by a note from internal finance seeking to make the date of priority for spectrum allocation also "be the same as the date of . . . application" subject to eligibility (ibid: A637). The situation was compounded by a missive from MoF on November 22, 2007, seeking some clarifications, but ultimately, the original approval regarding issue of LoIs was reiterated on December 4, 2007 (ibid: A641). The LoIs were issued on December 10, 2007, to companies who had filed applications on or before September 25, 2007.

18 To be fair, the committee's brief was limited to examining TRAI's August 2007 recommendations. However, it is noteworthy that the committee did not examine the option to activate unified licensing in greater detail even after noticing that only two categories of licensing were available under NTP 1999 after the addendum was issued on November 11, 2003, while rejecting TRAI's recommendation to examine the issue and specify quantum of license fee for UAS license without spectrum. Instead, the committee suggested that the recommendation be referred back for the limited purpose of evolving a policy for spectrum reservation to incumbents; even this suggestion did not find acceptance with the commission.

19 The letter cannot be seen in isolation of an earlier set of correspondence during March 2007 and June 2007 between MoF and DoT (and the

cabinet secretariat) regarding the GoM established on February 23, 2006, after its terms of reference were changed in December 2006. MoF had indicated its view that spectrum pricing was of immense importance for the growth of telecom and needed to be discussed in the GoM, a view that did not prevail given the DoT's insistence that spectrum pricing was within its decision-making competence. The gist of the correspondence is given in Joint Parliamentary Committee (JPC 2013: 132–134).

20 This presentation is stated to have been "general in nature" and "covered the basic guiding principles for allocation of Radio Frequency Spectrum, procedure for spectrum allocation, existing criteria for additional spectrum, TRAI's recommendations, Telecom Engineering Centre's recommendations, etc." (JPC 2013: 81). In terms of sequencing, it followed the exchange of letters between the minister of communications and information technology and the prime minister on licensing and spectrum developments.

7

THE THIRD REGULATORY PHASE

7.1 The third phase: 2008 onwards

The dawn of 2008 – a year that saw the ramping up of competition in the telecom services market to internationally unprecedented levels – was preceded by one bit of correspondence within the executive branch that is important for the analysis of the events that followed. On December 26, 2007, the minister of communications and information technology wrote to the prime minister on "various issues like criteria for allotment of UAS [unified access service] licences, use of dual spectrum, issue of new licences, issue of LoIs [letters of intent], etc." (DoT 2011a: 76). This letter inter alia informed the prime minister that the Cellular Operators Association of India (COAI) had not cooperated with the Department of Telecommunications (DoT) in solving the subscriber linked criteria (SLC)–related problem for additional spectrum amicably and that the association wished to challenge the dual-technology policy decision, showing that "the attitude of COAI is to maintain their monopoly in the sector by avoiding healthy competition and level playing field" (ibid: A672). It was also brought out in the letter that the Telecom Regulatory Authority of India (TRAI)'s recommendations suggesting that no cap be placed on the number of access service providers in any service area was accepted by DoT on October 17, 2007, "in order to encourage more competition in the [t]elecom [s]ector" and that "this is the first time that December 2005 UASL guidelines are being implemented in letter and spirit in view of TRAI recommendation" (ibid: A674). The letter also stated that the first come first served (FCFS) policy for grant of UAS licenses was a three-stage process consisting of (1) issuance of LoI wherein "an application received first will be processed first and if found eligible will be granted LOI"; (2) issuance of license wherein FCFS was applicable, but given the decision not to cap the number of UAS licenses leading to the proposal to issue a large number of LoIs simultaneously, FCFS would operate such that "an applicant who fulfils the conditions of LOI first will be granted licence first, although several applicants will be issued LOI simultaneously"; and (3) grant of wireless license, an independent license for allotment of radio frequency authorizing launch of Group Speciale Mobile (GSM)/Code Division Multiple Access (CDMA)-based mobile services, wherein again FCFS policy would be applicable (ibid: A675).

The tweaking of the FCFS policy was ostensibly due to the simultaneous issuance of LoIs; the simultaneity itself resulted from the imposition of a cut-off date for processing applications received until that date. This nuance becomes important since it was the policy prior to the simultaneous issuance of LoIs to maintain the seniority of applicants with effect from the date of application even though the companies were allowed to demonstrate their eligibility until such time as the LoI was issued. It has also been admitted that in the earlier cases, an "applicant company has also been permitted additional time to fulfill the requirement of LOI on their request beyond stipulated time mentioned in the LOI" (ibid: A646). In effect, prior to the decision taken to keep all pending applications on hold once TRAI's recommendations were sought on April 13, 2007, the processing of one case (or a few cases) at a time allowed DoT to maintain the seniority of applications for operating the FCFS policy according to the date on which the application was first filed, even for the next stages, such as signing of license agreement and issuance of wireless license after the LoIs were issued and complied with. This procedure was to be modified such that when a large number of LoIs were going to be issued simultaneously (there were 232 applications from 22 different companies received up to September 25, 2007), instead of maintaining the seniority according to the date the application was first filed, the company that fulfilled the LoI conditions first would be granted the license first. The simultaneous processing of applications also entailed that the eligibility of applicants needed to be reckoned as on the date of application and that no additional time could be granted to rectify the applications as hitherto. On January 7, 2008, DoT moved to formally obtain the approval of the minister of communications and information technology for the approval of the FCFS policy whose broad contours were as stated earlier. The secretary of DoT also proposed that a press release should also be issued in the matter. While approving the proposal based on the policy enunciated in his letter to the prime minister on the same date, the minister directed that the solicitor general's opinion could also be obtained (ibid: A649).

The solicitor general inter alia observed that a press release "makes for transparency" (ibid: A650), and an amended press release – in which the last paragraph that stated that "if more than one applicant complies with LOI condition on the same, the inter-se seniority would be decided by the date of application" was deleted on the grounds that it was a new stipulation (ibid) – came to be issued on January 10, 2008, notifying that DoT had accepted TRAI's recommendation not to cap the number of access service providers in any service area and had decided to "issue LOI to all eligible applicants on the date of application who applied up-to 25–9–2007" (ibid: A678). The press release also stated that the "UAS licence in broader terms is an umbrella licence and does not automatically authorize UAS licencees usage of spectrum to rollout [m]obile (GSM and/or CDMA) services" for which the UAS licensee had to "obtain another licence, i.e., Wireless Operating Licence which is granted on first-come-first-serve[d] basis subject to availability of spectrum in particular service area", and that DoT had been implementing an FCFS policy for granting UAS licenses "under which initially an application which is received first will be processed first and thereafter if found eligible will be granted LOI and then who so ever complies with the conditions of LOI first will be granted UAS licence" (ibid).

The press release of January 10, 2008, set the stage for what has been termed "a game-changing event when new licences were offered by [DoT] at a price equal to that discovered in the auction for the fourth cellular licence in 2001", taking the number of players in each service area to between 11 and 13 (Prasad 2011: 23). Of the 232 applications received up to September 25, 2007, "16 eligible applicant companies have to be issued 121 LOIs in total whereas 6 applicant companies have to be issued response of DoT" (DoT 2011a: A651). LoIs were issued on January 10, 2008, itself, and most companies complied with LoI conditions on the same day. It also transpired that some applicant companies challenged the FCFS policy contained in the press release before the Telecom Disputes Settlement and Appellate Tribunal (TDSAT), while others challenged the decision to grant LoIs in the Delhi High Court. Notwithstanding the legal challenge, DoT entered into license agreements for unified access services with compliant LoI holders – "with only one LOI holder in each service area in order of priority in one day"[1] (ibid: A657) – between February 27, 2008, and March 7, 2008.[2] The effective date of all licenses was January 25, 2008, and they were subject to the outcome of court cases.

The introduction of additional licensees in the access services market has been credited with reducing the market power of service providers but has also been blamed for creating "a high cost structure due to spectrum fragmentation" (Prasad 2011: 26). It has been suggested that the alternative of bringing in two or three additional operators, along with either indexing the price of start-up spectrum to technological improvements or auctioning it to new entrants (and auctioning the remainder to all comers holding license), and finally taking steps to promote contestability, "like the facilitation of spectrum sharing and incentives for mobile virtual network operators", would have worked better (ibid). While this alternative would arguably have had efficiency-inducing effects on the sector, what is more pertinent is whether the structural conditions that would have enabled this or any other alternative to receive policy attention were in place in the telecom sector in the period leading up to 2008. Without foregoing the resolve to steer clear of any alleged criminal wrongdoing, it can still be argued in the shadow of the TRAI Act and the Indian Telegraph Act (ITA) that what occurred was almost inevitable given the slew of licensing decisions taken by DoT in exercise of its discretionary powers in a reactive fashion over the years – certainly from permitting limited mobility in 2001 if not from the first time contractual conditions were observed in the breach at the time of the migration package in 1999. The non-implementation of unified licensing in the years since 2003 was of course the proximate reason spectrum was not considered independently of grant of license in 2007–8, but the ultimate responsibility for the hurried introduction of hypercompetition in the sector must be laid at the door of the institutional imbalance between licensor and regulator on licensing and spectrum-related matters. The moral hazard inherent in making an expert body only recommendatory on vital regulatory issues – alluded to earlier – compounded the misplaced optimism in expecting the licensor to follow the regulator's advice in letter and spirit. Neither party had the incentive to chart a steady and consistent course to regulatory certainty and the consideration of alternatives in an atmosphere infused with expertise and mutual institutional regard.

The third generation (3G) and broadband wireless access (BWA) auctions

The uneven relationship between TRAI and DoT was also reflected in the events leading up to the auction of 3G spectrum in 2010. After the issuance of UAS licenses in January 2008 was out of the way,[3] DoT wrote to TRAI on April 10, 2008, seeking reconsideration of its recommendations on this issue and stating that the ministry was considering the inclusion of other Indian operators or foreign operators in joint venture with Indian companies and who had at least one year's experience in telecom service provisioning (of providing at least half a million wireless access lines) in the spectrum allocation process. This reference elicited the regulator's response on April 25, 2008, in accordance with the statutory provision regarding responding to a request for reconsideration within 15 days (TRAI 2008a).

In its response, TRAI indicated that it had deliberated upon the option of treating existing UAS and cellular mobile telephone service (CMTS) licensees "as well as non telecom companies whether Indian or foreign . . . as potential candidates for offering 3G services in India" at the time of framing its September 27, 2006, recommendations but did not consider it appropriate, feasible, or viable (ibid: 3). It reiterated the reasons behind the recommendation: (1) the primary importance of affordability of new services; (2) the low incremental cost at which existing operators could supply 3G services given their "huge investment in the infrastructure"; (3) speed of rollout, which was likely to be higher for existing operators; (4) existence of adequate level of competition; (5) inadequate spectrum availability, with advantage likely to accrue from deploying 3G as an overlay network; and (6) the absence of a clear roadmap for spectrum availability (ibid: 3–5). TRAI also asserted that the Indian market – while on a high growth path – "is primarily voice centric and is likely to remain unchanged for some time" in support of its recommendation to auction 3G spectrum in lots of 2 × 5 MHz (ibid: 7). It referred to DoT's rejection of the category of "UAS licensee not wishing to utilize spectrum" to argue that a successful new entrant could not be accommodated in the existing UAS license category without becoming eligible for start-up 2G spectrum as well and that granting UAS license automatically would "create legal complications and litigations" as well as be construed as "an indirect backdoor entry" (ibid: 8). It buttressed its argument by stating its apprehension that outsiders proposing to enter the Indian telecom market could drive up 3G spectrum prices resulting in high tariffs and would also "deprive the existing operators from providing . . . 3G services" (ibid); it also viewed the imposition of the experience condition on prospective new entrants as upsetting the level playing field. On these considerations, TRAI reiterated that (1) the auction should be restricted to existing licensees; (2) the 'spectrum module' for auction should be 2 × 5 MHz; (3) simultaneous ascending auction as recommended in September 2007 may be accepted; (4) the total availability of spectrum should be publicized at the time of auction, and those who do not obtain auction in the first phase should be placed in queue and "allotted spectrum as and when available on the same terms and condition[s] as granted to the licensees in the first phase"; and (5) the situation should be reviewed after three years based on technological advancements (ibid: 9–10). TRAI thus stuck to its position that 3G spectrum should be allocated by auction to existing access service providers.

The set of TRAI's recommendations on 3G and BWA auctions, given between 2008 and 2009, were marked by an intense exchange of views between licensor and regulator regarding the spectrum auctions. They were followed by more than a year's worth of planning and implementing the auction process for 3G and BWA spectrum within DoT that is not of much relevance for the present narrative. All that can be said about the process is that the regulator played a secondary role in the auction proceedings related to a resource that is arguably the most valuable for the provision of mobile access services. While TRAI did try to make itself heard, it was ultimately DoT – with its stranglehold over information regarding *available spectrum* – that determined the major parameters of the first round of spectrum auctions that could authentically be labelled as such. Whether on publicizing the details of available spectrum, or on limiting the eligibility to existing operators, or on a specific frequency band (for BWA), or on the number and size of blocks to be auctioned, or on reserve prices, or on tenure of spectrum usage, or on recurring spectrum usage charges (SUC), or even on tweaking the licensing policy to provide (or desist from providing) for new categories of UAS licensing, it was DoT that throughout dictated the terms of the exchange while TRAI remained reactive in its regulatory approach. This imbalance was a reflection of the statutory weakness inherent in section 11 (1) (a) (viii), which gave limited purchase to TRAI on spectrum management; at the same time, it was also indicative of the secondary institutional role TRAI had come to play in the sector. Even though clauses (i) and (ii) of that section gave it a slightly more enhanced recommendatory role regarding licensing new entrants and determining the terms and conditions of license, the fifth proviso that defined the procedure to be adopted in case the government did not agree with the regulator's recommendations on these issues proved unequal to the task of protecting the expert regulatory recommendations against the sovereign will and/or caprice of the licensor. The weakness in the statute was compounded by TRAI's inability to impose its expertise on DoT on the few occasions it tried to do so. The scope and direction of the 3G and BWA auctions were determined largely at the instance of DoT, with TRAI acceding – at times dutifully and at others with a show of resentment.

The auctions themselves came to be conducted in April and May 2010.[4] They were to be simultaneous ascending e-auctions conducted over the internet.[5] By stipulating that the award of spectrum was not intended to confer rights to provide services, the Notice Inviting Applications (NIA) effectively delinked spectrum from license for the first time; it was necessary for the winner to possess or acquire a UAS or CMTS license in the relevant service area to provide licensed services in the case of 3G spectrum (or, in addition, an ISP license for BWA services). By keeping the auctions open to new entrants in both 3G and BWA, the NIA sealed the government's keenness to induct additional competition that it had signalled in its information memorandum issued in October 2009 listing the steps being taken to "provide new entrants with the opportunity for the spectrum being auctioned and subsequently roll out their networks on a more equal footing with existing operators" (DoT 2009b: 1). While prospective new entrants were required to obtain a UAS license prior to providing services, the NIA made it clear that this did not imply that "such Bidders, if successful in winning 3G/BWA Spectrum and in obtaining a UAS licence, will also be awarded start-up spectrum" in 2G (DoT 2010a: 13). The NIA also specified the applicable SUC regime that – while not adding the 3G or BWA spectrum for computing the slab rate – made revenue from 3G services part of the adjusted gross revenue (AGR) for existing licensees, stipulated a charge of 1 per cent of AGR from BWA services irrespective of the license held, and also stipulated that the slab rate for standalone 3G operators shall be equal to the lowest slab rate (3 per cent), with a moratorium of one year. On offer were three 3G blocks each of 2 × 5 MHz in 17 service areas, four 3G blocks each of 2 × 5 MHz in 5 service areas, and 2 unpaired BWA blocks of 20 MHz each in all 22 service areas. One block each of 3G and BWA spectrum was reserved for the public sector Bharat Sanchar Nigam Limited (BSNL) and Mahanagar Telephone Nigam Limited (MTNL) in their areas of service, with the provision that they would be required to pay the winning price achieved in the respective service area for each category.

The auctions attracted wide participation. The final list of bidders included nine companies for 3G spectrum and 11 companies for BWA spectrum. The 3G auction went on for 183 rounds over 34 days from April 9, 2010, to May 19, 2010, and all the 71 blocks that were put up for auction were sold. The BWA auction went on for 117 rounds over 16 days from May 24, 2010, to June 11, 2010, and all 44 blocks on auction were sold. Table 7.1 indicates the winning bids in both auctions. Together, they yielded over INR 106,262 crore to the government after including the matching price paid by BSNL and MTNL.

Table 7.1 Results of 3G and BWA auctions

(All prices, revenues in INR crore)

Service area **(Reserve Price 3G/BWA)**	*Blocks*		*Winners*		*Winning price*		*Total revenue*	
	3G	*BWA*	*3G*	*BWA*	*3G*	*BWA*	*3G*	*BWA*
Delhi (320/160)	3	2	Vodafone, Bharti, Reliance	Infotel, Qualcomm	3,316.93	2,241.02	9,950.79	4,482.04
Mumbai (320/160)	3	2	Reliance, Vodafone, Bharti	Infotel, Qualcomm	3,247.07	2,292.95	9,741.21	4,585.90
Maharashtra (320/160)	3	2	Tata, Idea, Vodafone	Bharti, Infotel	1,257.82	915.64	3,773.46	1,831.28
Gujarat (320/160)	3	2	Tata, Vodafone, Idea	Infotel, Tikona	1,076.06	613.85	3,228.18	1,227.70
Andhra Pradesh (320/160)	3	2	Bharti, Idea, Aircel	Aircel, Infotel	1,373.14	1,059.12	4,119.42	2,118.24
Karnataka (320/160)	3	2	Tata, Aircel, Bharti	Infotel, Bharti	1,579.91	1,543.25	4,739.73	3,086.50
Tamil Nadu (320/160)	3	2	Bharti, Vodafone, Aircel	Infotel, Aircel	1,464.94	2,069.45	4,394.82	4,138.90
Kolkata (120/60)	3	2	Vodafone, Aircel, Reliance	Infotel, Bharti	544.26	523.20	1,632.78	1,046.40
Kerala (120/60)	3	2	Idea, Tata, Aircel	Infotel, Qualcomm	312.48	258.67	937.44	517.34
Punjab (120/60)	4	2	Idea, Reliance, Tata, Aircel	Bharti, Infotel	322.01	332.27	1,288.04	664.54
Haryana (120/60)	3	2	Idea, Tata, Vodafone	Infotel, Qualcomm	222.58	119.90	667.74	239.80
Uttar Pradesh (E) (120/60)	3	2	Aircel, Idea, Vodafone	Tikona, Infotel	364.57	142.50	1,093.71	285.00
Uttar Pradesh (W) (120/60)	3	2	Bharti, Idea, Tata	Tikona, Infotel	514.04	183.87	1,542.12	367.74
Rajasthan (120/60)	3	2	Reliance, Bharti, Tata	Tikona, Infotel	321.03	97.32	963.09	194.64
Madhya Pradesh (120/60)	3	2	Idea, Reliance, Tata	Augere, Infotel	258.36	124.66	775.08	249.32
West Bengal (120/60)	4	2	Bharti, Reliance, Vodafone, Aircel	Infotel, Aircel	123.63	70.97	494.52	141.94
Himachal Pradesh (30/15)	4	2	Bharti, STel, Idea, Reliance	Tikona, Infotel	37.23	20.66	148.92	41.32
Bihar (30/15)	4	2	STel, Bharti, Reliance, Aircel	Infotel, Aircel	203.46	99.28	813.84	198.56
Orissa (30/15)	3	2	STel, Aircel, Reliance	Aircel, Infotel	96.98	63.63	290.94	127.26
Assam (30/15)	3	2	Reliance, Bharti, Aircel	Aircel, Infotel	41.48	33.02	124.44	66.04
North East (30/15)	3	2	Aircel, Bharti, Reliance	Infotel, Aircel	42.30	21.27	126.90	42.54

(Continued)

Table 7.1 (Continued)

(All prices, revenues in INR crore)

Service area **(Reserve Price 3G/BWA)**	*Blocks*		*Winners*		*Winning price*		*Total revenue*	
	3G	*BWA*	*3G*	*BWA*	*3G*	*BWA*	*3G*	*BWA*
Jammu and Kashmir (30/15)	4	2	Idea, Aircel, Reliance, Bharti	Infotel, Aircel	30.30	21.27	121.20	42.54
All (3,500/ 1,750)	**71**	**44**			**16,750.58**	**12,847.77**	**50,968.37**	**25,695.54**
BSNL/MTNL	22	22					16,750.58	12,847.77
Grand total	93	66					**67,718.95**	**38,543.31**

Source: www.dot.gov.in

It has been noted that the final pan-India 3G and BWA bid amounts were five and seven times the respective reserve prices and that "on an average, the BWA bid amount in each circle was about 68 per cent of the corresponding 3G bid price" (Sridhar 2012: 105), even though the BWA reserve price was half that of 3G and the start-up spectrum was double that of 3G. The success of the 3G auction, which far exceeded revenue expectations, has been attributed to a combination of factors including risk-loving bidder behaviour that led to "a greater premium on the regret of losing over the joy of winning" (Kumar 2011: 128). However, it has also been noted that it was "unable to achieve the objective of attracting new entrants to encourage stronger competition with incumbents" since most licenses were bagged by entrenched 2G incumbents (ibid).

The comptroller and auditor general (CAG)'s report

The trigger for the performance audit report sent to DoT in November 2010 was the issuance of "120 new licences for unified access services on the same day" in January 2008, at a "price which had been discovered in 2001" (CAG 2010: iii). The report noted that questions had been raised regarding transparency in the allocation process and the failure to maximize revenue generation from spectrum allocation given that it was a national asset. Claims that ineligible applicants were given licenses "at a price which appeared far below what has been perceived to be the appropriate market price in 2008" was cited as sufficient justification "to review the entire process of issuance of licences, award of spectrum and the implementation of the UAS regime" while accepting the government's prerogative to formulate policy (ibid). The major findings of the audit report were: (1) that there were gaps in policy implementation, especially in not implementing the second phase of policy approved by the cabinet in 2003 in its decision to move to unified licensing and in DoT not associating the Ministry of Finance (MoF) in the efficient allocation and pricing of spectrum, leading to "issuance of licences in 2008 along with allocation of spectrum . . . by DoT at prices determined in 2001 which were based on a totally nascent market despite the sector witnessing substantial transformation and manifold growth" (ibid: v); (2) that the Telecom Commission was not consulted on important matters; (3) that MoF's views and concerns were overruled by DoT; (4) that the advice of the Ministry of Law and Justice was ignored; (5) that the prime minister's suggestions were not followed; (6) that DoT made arbitrary changes to the cut-off date in 2007; (7) that the FCFS policy was not followed; (8) that licenses were issued to ineligible applicants; (8) that "the presumptive value of 2G spectrum on account of grant of 157 licenses in different circles during 2007–08 would be in the range of approximately [INR] 58,000 crores to [INR] 1,52,038 crores" (ibid: ix); and (9) that DoT allotted spectrum to existing operators beyond contracted limits without imposing any upfront charges and the value of spectrum held by these operators at 2001 rates "would be in the range of [INR] 12,000 crores and [INR] 37,000 crores" (ibid). The highest of the three estimates[6] of the presumptive loss caused by spectrum allocation to 122 UAS licensees, 35 crossover licensees, and 13 operators holding 51 licenses given spectrum beyond 6.2 MHz without charging upfront spectrum fee calculated by CAG amounted to INR 176,645 crore.

The details of calculations and the particulars of the many deficiencies discovered by the CAG report need not detain us. We must note, however, the following aspects that crucially influenced the CAG in formulating its estimates of presumptive loss: (1) the notion of 'contractual spectrum' varying between 2 × 4.4 MHz GSM spectrum prior to 2001 and not more than 6.2 MHz and 5 MHz for GSM and CDMA spectrum thereafter (ibid: 8); (2) the observation that "generation of revenue for the Government is not within the scope of [TRAI's] mandate and hence not perhaps a basis for framing its recommendations" and that "protecting the financial interests of the Government should have been an important consideration for the DoT" while accepting the regulator's recommendations, more so when MoF was left out of the decision-making process (ibid: 16); (3) that all agencies of government (DoT, TRAI, MoF) had full knowledge of the scarcity and underpricing of spectrum but the entry fee of licenses continued to be pegged at 2001 rates even in 2007 "without delinking and independently discovering the price of spectrum through a market mechanism" (ibid: 17); (4) that there was no doubt the 2001 price "did not reflect the true economic value of a licence and the spectrum bundled with it in 2008" and most crucially that "since no price discovery of spectrum was attempted for 2G spectrum separately, *the entry fee discovered in 2001 is mainly the price of spectrum that came with UAS licence*" (ibid: 51, emphasis added); and (5) that the role of TRAI "would . . . appear to have been reduced to that of a hapless spectator as its recommendations were either ignored or applied selectively" (ibid: 58).

The composite nature of the spectrum problem in the context of the artificial policy division of access service market into fixed and cellular and the issues generated by the contradiction need to be firmly grasped to avoid analytical confusion that views the 2001 tenders as having yielded an upfront spectrum price. It was lock-in from previous tenders that was responsible for the incorporation of start-up spectrum in 2001 to ensure a level playing field; just as in the case of fixed service provider (FSP) tenders, the fourth CMSP tenders also did not include any upfront payment for spectrum as part of the entry fee, and no scarcity or market value of spectrum was determined by the tendering process. Only SUC needed to be paid in lieu of assured supply, and this was because no upfront price was being charged that the FCFS policy came to be imported into the cellular access licenses by virtue of their issuance on a 'continuous basis' a la FSPs. SLC and additional SUC for higher spectrum allocation were devices used by DoT to exercise executive discretion in spectrum allocation, and these too support the proposition that no upfront fee for spectrum was envisaged prior to the 3G and BWA auctions. Finally, the stipulation that no additional spectrum would be allotted on migration to the UAS license even after the payment of the difference in entry fee paid by FSPs and the fourth CMSP is reflective of the regulatory reluctance to charge an upfront fee, however arguable that position.

It was therefore an imperfect appreciation of legacy issues in the long history of the spectrum-license problem in India that led CAG to conclude that the entry fee discovered in 2001 was mainly the price of spectrum that came with UAS license. CAG's assertion that TRAI was a hapless spectator also needs to be tempered with a realistic understanding of statutory provisions that built moral hazard into its recommendatory role, which was in any case diminished after the amendment. Whatever the flaws of the non-evaluation of the scarcity value of spectrum (and we have listed some of these in our earlier discussion), it is a bit of an epistemic leap on the part of CAG to state that the 2001 tenders had yielded the price of spectrum that came along with the license. The regime – while arguably imperfect, both prone to discretionary excess as well as allowing the spectrum price to be loaded on to retail tariffs as variable cost – was based on SUC and nowhere envisaged even a modicum of change in property rights that undoubtedly would have accrued from a dispensation centred on upfront charges. CAG's leap of faith cannot be sustained on these grounds alone, whatever the merits of all its other observations, many of which are impeccable (and some the subjects of criminal cases and inquiries). CAG's presumptive valuation of spectrum based on a less than perfect understanding of legacy issues had an important influence on the way subsequent discourse regarding spectrum came to be conducted in the country.

Before we deal with this aspect of the 2G litigation, however, it is necessary to take note of a voluminous set of recommendations made by TRAI in May 2010 that have a bearing on spectrum valuation in the third phase of regulatory governance. TRAI's recommendations – made on May 11, 2010, even as the 3G and BWA auctions were in progress – were marked both by their volume and the web of linkages between the different parts that made up a complex structure requiring a close and careful reading of the interconnections to make sense of the whole.[7] The imprint of legacy issues is abundantly visible in these recommendations; the complexity of the recommended scheme is magnified if the regulatory contortions to get around some of them are closely examined, as is the exertion to instantiate the mythical level playing field. It is also true that they were made in the context of the furore generated by the controversy surrounding the issuance of UAS licenses in early 2008 and the extensive litigation it engendered. Nevertheless, the general direction with regard to spectrum was clear: that license and spectrum needed to be decoupled, that the upfront value of spectrum must be discovered from the market, and that SUC charges needed to be continued. While these were no doubt welcome assertions from a regulatory standpoint, what is of particular interest is the proposition that the prices of 2G spectrum bands were in a mathematical relationship with the auction prices that were being discovered in the 3G auctions. TRAI recognized that there were problems with adopting the 3G price as the current price of 2G spectrum in the 1800 MHz band; it noted that there were conflicting views about the applicability of the price of the 2100 MHz band to the 1800 MHz band and indicated it could not arrive at a definitive conclusion. TRAI stated that it was "separately initiating an exercise to further study this subject and would apprise the Government of its findings" and that "pending further deliberations . . . the 3G [p]rice may be adopted as the price for 2G" (ibid: 189).[8] On the argument that detailed studies had established the benefits in capital and operational expenditure of the 900 MHz band over the 2100 MHz band, and given the interim equivalence of the 2100 MHz and 1800 MHz bands, TRAI also "concluded that the value of 900 MHz band is at least 1.5 times that of 1800 MHz band" and that this principle could be applied for the 800 MHz band spectrum also (ibid: 192). This mathematical equivalence that TRAI attempted to establish not just between spectrum efficiency across bands but *also between spectrum prices without reference to the respective demand and supply in various bands* was the arguable regulatory innovation CAG found useful in its own estimation of loss from the allocation of licenses in 2008, once it had concluded that the price tendered in 2001 was for the upfront value of spectrum and therefore needed to be compared with more up-to-date upfront valuations of the same or similar goods. This background study will examine the judgments of the Supreme Court in the 2G cases specifically to analyse the treatment of the legacy and regulatory issues surrounding the scarce spectrum resource in the judicial discourse.

The 2G case and the presidential reference

The term '2G case' refers to a clutch of cases before the Supreme Court: (1) first, the common judgment written on February 2, 2012 by Justice G S Singhvi (on behalf of himself and Justice Asok Kumar Ganguly) of the Supreme Court in writ petition (civil) no. 423 of 2010 titled *Centre for Public Interest Litigation and Ors.* v. *Union of India and Ors.* (2012) 3 SCC 1 and another writ petition (no. 10 of 2011) titled *Dr Subramanian Swamy* v. *Union of India and Ors.* (the lead judgment); (2) second, the Supreme Court's order dated February 2, 2012, in a Civil Appeal (no. 10660 of 2012) titled *CPIL and Ors.* v. *Union of India and Ors.*; and (3) its judgments dated January 31, 2012, in *Dr Subramanian Swamy* v. *Dr Manmohan Singh and Anr.* (2012) 3 SCC 64. The last two sets were also rendered from the same bench consisting of Justices Singhvi and Ganguly. The 2G case led to a reference to the Supreme Court from the president that resulted in the apex court's opinions (from a five-judge bench) in *In re: Special Reference No. 1 of 2012* (2012) 10 SCC 1, which was delivered on September 27, 2012.

We propose to examine the import of the Supreme Court's pronouncements in the 2G case as well as its subsequent opinions (rendered by Justice D K Jain on behalf of four justices including the chief justice of India and supplemented by a separate, concurring opinion of Justice J S Khehar) in light of the licensing legacy, leaving out entirely their fallout on ongoing criminal prosecutions[9] as well on the investments made by companies whose licenses have been cancelled by virtue of the lead judgment.[10] The focus will be on governance arrangements for spectrum and its regulation and how this has been affected by the judicial pronouncements. As such, even the reading of the judgments and opinions will be restricted to explicating the implications of the Supreme Court's construal of scarce natural resources on electromagnetic spectrum in light of the ambiguous relationship noted between spectrum and license through the history of privatization of telecom service provisioning in India. The lead judgment and the opinions rendered on the president's reference will therefore engage our exclusive attention. The order in Civil Appeal no. 10660 of 2010 –directions regarding the monitoring of police investigations – and the judgments in *Dr Subramanian Swamy* v. *Dr Manmohan Singh and Anr.* (2012) 3 SCC 64 – pronouncements related to the Prevention of Corruption Act, 1988, and grant of prosecution sanction under that statute – are not considered since they are not strictly relevant to the present analysis.

i The lead judgment: *Centre for Public Interest Litigation and Ors.* v. *Union of India and Ors.* (2012) 3 SCC 1:

> In its judgment, the Supreme Court took up the following questions it termed "important" that arose for consideration from the petitions in the case: (1) whether the government has "the right to alienate, transfer, or distribute natural resources/national assets otherwise than by following a fair and transparent method consistent with the fundamentals of the equity clause enshrined in the Constitution"? (2) whether TRAI's recommendations of August 28, 2007, "for grant of Unified Access Service Licence . . . with 2G spectrum in 800, 900 and 1800 MHz at the price fixed in 2001, which were approved by [DoT], contrary to the decision taken by the Council of Ministers on 31.10.2003"? (3) whether the process of DoT's grant of UAS licenses during the period September 2007–March 2008 "to the private respondents in terms of the recommendations made by TRAI is vitiated due to arbitrariness and malafides and is contrary to public interest"? (4) whether the FCFS policy followed by DoT for grant of licenses "is *ultra vires* the provisions of Article 14 of the Constitution and whether the said principle was arbitrarily changed" by the Minister without consulting TRAI, with a view to favour some of the applicants? (5) whether the licenses granted to ineligible applicants and those who failed to fulfil the terms and conditions of license "are liable to be quashed"?
>
> Before proceeding to the substantive analysis of the arguments, the judgment observed that it did not consider it proper to refer to either the findings in the CAG report (since it was under examination by Parliamentary Committees) or the one-man committee report (on the grounds that the government "has already taken a decision to segregate spectrum from licence and allot the same by auction"). It also quoted from a press statement issued by the "present Minister of C&IT" on January 29, 2011, that laid out the new policy dispensation (paragraph 61).

The judgment allowed the writ petitions and ordered that: (1) the licenses granted on or after January 10, 2008, following the press releases of that date were illegal and were quashed; (2) this decision would become operative after four months; (3) TRAI shall make fresh recommendations for grant of licenses and allocation of spectrum in the 2G band in the 22 service areas by auction keeping in view the government's policy decision of 2011; (4) the government shall consider and decide upon TRAI's recommendations within one month and fresh licenses shall be "granted by auction"; and (5) private respondents who benefited from DoT's action shall pay costs as stipulated. The court also made it clear that its observations in this judgment shall not affect pending criminal investigations or cause prejudice to those facing prosecution.

ii The opinions: *In re: Special Reference No. 1 of 2012* (2012) 10 SCC 1:

a The majority opinion:

The majority opinion reproduced the full text of the reference made under Article 143 (1) of the constitution to the Supreme Court by the president of India on April 12, 2012. After briefly setting out the history of telecom licensing and spectrum allocation in the country prior to 2008 and some subsequent developments such as the allocation of 3G spectrum by auction and the government's decisions on some of the TRAI recommendations of May 11, 2010, the reference cited the Supreme Court's judgment in the 2G case and its subsequent interlocutory application seeking clarification and a review petition in the matter.

After concluding that the reference was maintainable, the opinion first considered what was "precisely stated in the 2G Case and decipher the law declared in that case" (paragraph 64). The opinion concluded that the judgment in the 2G case did not deal with modes of allocation of natural resources other than spectrum.

The opinion "unhesitatingly" answered the question of whether auction as a method of disposal of natural resources was to be declared a constitutional mandate under Article 14 "in the negative" on various grounds.

> The opinion concluded that "auctions are not the only permissible method for disposal of all natural resources across all sectors and in all circumstances" (paragraph 150). It "respectfully" declined to answer the remaining questions since the answer "would have a direct bearing on the mode of alienation of [s]pectrum" in view of the Attorney General's statement that the government was not questioning the correctness of the judgment in the 2G case (paragraph 151).

b The concurring opinion:

> In his separate opinion, Justice Khehar endorsed the main opinion; the concurring opinion stated that while auction is "certainly not" a constitutional mandate, "it can surely be applied in some situations to maximize revenue return, to satisfy legal and constitutional requirements" (paragraph 12). It declared that "there cannot be a dissipation of material resources free of cost or at a consideration lower than their actual worth" and concluded by observing that "one set of citizens cannot prosper at the cost of another set of citizens, for that would not be fair or reasonable" (paragraph 13).

A judgment and an opinion: the future of regulatory governance in telecom

It has been remarked that the judgment in the 2G case "has the widest material impact with no substantive legal reasons/support" and that the opinion on the presidential reference "restores substantive reasons but with no impact on 2G-spectrum allocation" (Krishnaswamy 2013: 30).

The judgment in the 2G case appears thus to lend itself easily to legalistic critique of the way it strings together bits and pieces of various doctrines and reads into the equality clause of the constitution one particular economic philosophy that elides public auctions with public interest. The observations in the opinions on the reference that Article 14 does not present any economic policy as a constitutional mandate and that Article 39 (b) imposes no restrictions on the means adopted to subserve the public good leading to the unambiguous pronouncement that "auctions are not the only permissible method for disposal of all natural resources across all sectors and in all circumstances" (paragraph 150) effectively overrules the assertion in the 2G case judgment that "while transferring or alienating natural resources, the State is duty bound to adopt the method of auction by giving wide publicity so that all eligible persons can participate in the process" (paragraph 76). The opinion made it clear that neither the fundamental right to equality (Article 14) nor the directive principle of state policy embodied in Article 39 (b) could be utilized to support the proposition that auction was the only permissible method of disposal or distribution of natural resources under all circumstances, and it was the executive's decision – conveyed to the Supreme Court during the hearing of the presidential reference – to effectuate a change in policy with regard to spectrum allocation that foreclosed the possibility of the court recording its opinion regarding the remaining three questions related to 'alienation' of spectrum. The government's decision in its executive capacity to alter licensing and spectrum policy on January 29, 2011, even as the 2G case was under adjudication in the Supreme Court (and extracted by it and taken judicial notice of at paragraph 61 of its lead judgment) was responsible for spectrum being kept out of the advisory ambit of the Supreme Court rather more than the court's reticence or reluctance to address the issue of spectrum allocation. That the opinion has had the effect of overruling the *ratio decidendi* of the 2G case judgment is trite, and except for the government's decision to change its policy, there was no reason the ratio laid down in the opinion could not have applied to spectrum, especially since the main opinion recorded its view that the judgment in the 2G case had not considered the case of auction in general, but specifically evaluated "the validity of those methods adopted in the distribution of spectrum from September 2007 to March 2008" (paragraph 78). If the judicial process had been allowed to run its course without the executive deciding to make a change in policy, it is possible that the spectrum allocation policy would not have remained tied down by the ratio in the 2G case judgment until it was legally tested sometime in the future, even though the decision in the *lis* that led to the cancellation of licenses would still have been left undisturbed.

The 2G case judgment is remarkable for its certitude regarding the 2007–8 DoT exercise that it termed "wholly arbitrary, capricious and contrary to public interest", which surely shaped its conclusion that it violated the constitutional doctrine of equality as well (paragraph 77). It would appear that the anxiety to correct that unsustainable executive action overshadowed the need to meticulously lay down sustainable legal arguments and public reasons for the decision. That the executive decided to make changes to licensing policy is a measure not only of political imperatives arising from public opinion and so on, but also of the burden imposed by a concurrent examination of constitutional and criminal issues by the apex court during the litigation. It is also not beyond speculation that the announcement of NTP 2012 – with its emphasis on delinking spectrum from license on May 31, 2012, soon after the judgment – was influenced by the tenor of the judgment in the 2G case.

This brings us to the question of the implications of the judgment and the opinion on the future of regulatory governance in the telecom sector. Since the decision in the 2G case remains valid for lack of contestation, its terms need closer examination to discern their effects on the institutional environment. The first implication is clear enough: spectrum auctions will be mandatory for the immediate future until the policy in this regard is taken up for modification and such policy change survives legal challenge. Given that NTP 2012, which formalized "market related processes" for determining the price at which spectrum will be made available (NTP 2012: 12), is of recent vintage, it is unlikely for it to be changed by the executive in the near future. Since the opinion in the presidential reference desisted from answering the last three questions posed therein that related both to legacy issues, including spectrum 'giveaways' at different points in time, and to practical matters likely to arise in the future, such as the validity period of auction determined prices, ceilings on spectrum holding, and criteria for allocation in cases of inadequate competition for spectrum, it is now for the executive branch to grapple with these and other 'level playing field' issues going forward. Given that many of these issues are techno-economic in nature, it is obvious that TRAI would be called in to assist in the spectrum-related decision-making process more frequently than has hitherto been the case. This – while perhaps an unintended consequence – is a corrective to the legislative reluctance to assign a meaningful role to the regulator in spectrum management as evidenced by the unsatisfactory construction of section 11 (1) (a) (viii), which covers TRAI's recommendations for the efficient management of available spectrum. The contours of efficient management are now discernible and include making actionable recommendations on spectrum pricing that do not overlook "basic constitutional postulates and established principles" so that these recommendations would not "deny people from participating in the distribution of national wealth and benefit a handful of persons" (paragraph 74 of the 2G case judgment). The highest court's admonition that TRAI's recommendations of August 28, 2007, "were flawed in many respects" and that their implementation by DoT "resulted in gross violation of the objective of NTP 1999 and the decision taken by the Council of Ministers on 31.10.2003" (ibid) is likely to serve as a cautionary signal in both

TRAI's exercise of its expanded recommendatory jurisdiction as well as in DoT's treatment of the recommendations made by the regulator.

With all respect to the law articulated in the judgment, it must be stated that while the public trust doctrine is indeed applicable to spectrum, it would be misdirected to conceive of this doctrine as encouraging auctions as the preferred methodology for price discovery and allocation and thereby creating further incentives for vesting property rights in the winning bidders. The licensing regime for access licenses can be characterized as vesting quasi-property rights in the licensees with respect to spectrum given that – even though they do not own the spectrum – they retain a reasonable expectation of renewal of usage rights by virtue of the long-term nature of the contract. The quasi-property regime is supported by the instinct for liberalizing spectrum use not only in terms of permitting deployment of different technologies but also by calibrated easing of restrictions on sharing and trading of spectrum.[11] For all its drawbacks, the command and control approach adopted in the early years of licensing had not insisted upon upfront payment of spectrum value and went with a 'pay as you use' usage charge regime that dampened potential enthusiasm for vesting property rights in licensees. As argued, the access licensing contract regime was designed in a restrictive fashion with an eye on parting with the government's exclusive privilege under section 4 of ITA rather than with detailed attention to spectrum rights or use values up to and beyond the tenders in 2001, and this command and control dispensation was sufficient to cabin the loose property rights in the spectrum resource granted to licensees. With the present change in the policy stance occasioned by the 2G case judgment and aided by the exemption given to spectrum from the more robust articulation of constitutional provisions in the main advisory opinion, the luxury of an ill-defined property rights regime in spectrum is no longer available, either to the executive branch or to the fourth, regulatory branch. With the 2G case judgment ruling the spectrum field, it becomes mandatory for spectrum allocations to be done through auctions, which in turn entails that stronger property rights must be vested in whoever puts in the winning bids.[12] In fact, adequate reciprocal consideration for alienation of the spectrum resource to private parties cannot be extracted unless private property rights in the resource are gradually enhanced. The logic of credible regulatory commitment would imply that greater property rights be vested in investors' spectrum holdings in order to attract higher upfront prices. The loose, quasi-property rights regime in spectrum is thus set to be replaced. The replacement will naturally entail a more stable property rights regime that will work to the advantage of spectrum holders (licensees); it is trite to state that the advantage to entrenched licensees would be greater given the persuasive economic argument that incumbents not only hold large swathes of spectrum but are also better placed to corner additional tranches that the new policy promises to bring under the hammer for expanding commercial mobile services.

That the public trust doctrine has come to be used in support of this transitioning of spectrum rights to a more incumbent-friendly regime is indeed ironic. There was debate in the arguments during the presidential reference about whether the doctrine prohibits *any* alienation of natural resources and therefore should be applied only narrowly to preserve unimpeded public access to common resources such as air and water. The country of its origin has seen arguments for it to be extended to all natural resources including electromagnetic spectrum, with some parties going so far as arguing in an amicus brief that the government should be deprived of "the authority to sell off the public airwaves" (Ryan 2004: 337). Such creative arguments may not only not pass muster in the developing country context of India, but have also been noted to address only half the public trust doctrine's potential, because the doctrine "operates as a superior right guaranteeing qualified access to *all* kinds of property, whether it is owned privately, held by the state, or *unownable*, like the air and the sea" (ibid, emphasis in original). This chapter argues it is this second, superior right of qualified public access to natural resources that is mangled in the court-influenced policy change that seeks to fix auctions as the only mode of alienation of spectrum and thereby pave the way for the increasing vestment of property rights in licensed users, thus foreclosing the debate opened up by technological developments that seek to view spectrum as a commons rather than as private property. A stable property regime is no doubt a desirable regulatory objective; at the same time, plumping for a private property dispensation on the back of a narrow legal reading of the policy and regulatory situation is not supportive of an informed discussion of its relative merits over a commons regime[13] in as technologically oriented a sector as telecom. The irony is complete when we notice that incumbent service providers will have all the incentive to bid for and retain large spectrum holdings with increasing emphasis on the right to exclusive possession[14] in the name of the doctrine of public trust, even as arguments calling for harnessing technological development[15] to ensure unhindered access to spectrum as commons are gaining salience.

We turn next to the constitutional imperatives the judgment (and, for that matter, the advisory opinions) relies upon to support the case for spectrum auctions. Article 14 has been the primary vehicle for the articulation of the need for fair and just distribution of natural resources held in public trust. The judgment differs from the main opinion in that it views public auction as the only valid means for alienation, while the main opinion holds that there is no single method prescribed constitutionally for effecting the distribution of natural resources. Legal correctness of the respective articulations aside, it is somewhat obvious that both views are steeped in the private property rights tradition in its applicability to spectrum and do not consider the case for retaining the freedom of access to the public in the spectrum resource. Whatever the merits or demerits of this 'Demsetzian turn',[16] it appears disingenuous to draw support for a property rights paradigm from the doctrines of public trust or equality. Granted that what was at issue in the 2G case was the *lis* between parties that did not exist in an academic vacuum, but it is anachronistic that no arguments were made for the retention of a modicum of *public* nature of spectrum given the incentives for underutilization of large quantities of bandwidth endemic to private-property-based licensing regimes.[17] Persuasive arguments have also been made for regulating devices rather than the resource – drawing upon the analogy of the highway system (Lessig 1999; Ryan 2004) – for a more sensible regulatory dispensation, and these too would require the retention of public access to spectrum rather than increasing exclusive property rights vested in current holders.

The judgment is more open to criticism on the grounds that it reads into Article 14 a particular economic philosophy than the advisory opinions that interpret the constitutional provision in a broader manner. The judgment is premised on the argument that natural resources possess intrinsic utility to mankind and are amenable to distribution by the state, which must, however, ensure that constitutionalism is reflected at every stage of the distribution. It draws upon the Article 14 doctrine of equality (as well on the directive principle contained in Article 39 (b)) to note that "people [must] be granted equitable access to natural resources and/or its products and that they are adequately compensated for the transfer of the resource to the private domain" (paragraph 69 *supra*), but the judgment concentrates on the second aspect of adequate compensation for transfer to private domain to the neglect of the former aspect of ensuring equitable access. The focus of the judgment thereafter is on ensuring that constitutional principles are written into the procedures for transfer of rights in the natural resources to the private domain, and it is in this endeavour that it reads auctions as a constitutional command. This ratio has in effect been overridden by the advisory opinions that have read Article 14 (and Article 39 (b)) more capaciously such that auctions do not become a constitutional mandate, subject to the caveat that the method of distribution adopted must not fall foul of the judicial review requirements of executive action not being arbitrary, unfair, unreasonable, or capricious. But the main opinion too is open to the criticism that it could have delved deeper into the doctrine of public trust and laid down bright lines that could have guided future policy making, especially with regard to spectrum. The government's concession regarding spectrum no doubt left little advisory leeway for the apex court, but its admitted reluctance to interrogate the contours of the public trust doctrine has important implications for the future of the telecom services sector. While the opinion has made it clear that the Article 14 test would be applicable to executive policy decisions and the means adopted with regard to alienation of natural resources, whether the public trust doctrine is also applicable to such decisions to ensure a modicum of guarantee for public access to spectrum has been left open for future judicial determination. Until then, it would appear that the twin judicial determinations, along with the stated policy resolve to delink spectrum from license, would have the effect of locking in the sector to a private-property-rights based spectrum allocation regime focusing on resource regulation. While that may arguably be good from the point of view of encouraging stability in the regulatory regime and enhancing credible commitment for attracting further investment from entrenched and expectant corporate interests, whether it is the best alternative for the orderly growth of this technologically dynamic sector is a matter worth pondering.

7.2 Crafting order: summary and conclusion

The history of two decades of the regulatory institutions in the telecom services sector has been neither simple nor straightforward. The first stage was marked by bitter turf battles between DoT and TRAI, unsurprising given the long-held reluctance of the executive branch to give up regulatory powers. The combination of legislative ambiguity, executive intransigence, and judicial intervention brought the contradictions in the privatization process to a head, requiring resolution by recourse to a considerable reworking of the statutory framework for regulation.

The Amendment Act of 2000 carried through the categorization, initiated at the behest of the judiciary, of regulatory functions into adjudicatory and non-adjudicatory, with the first of these hived off to a specialized oversight tribunal. The non-adjudicatory functions, too, were divided by statute into those TRAI would give recommendations upon and those it would 'discharge'. The subtle downgrading of the independent regulatory authority (IRA) by virtue of the creation of TDSAT was compounded by the absence of any meaningful role for it in licensing matters, spectrum management, and equipment standards. The institutional imbalance between DoT and TRAI thus continued under the amended Act as well and caused many subsequent regulatory mishaps.

The inadequate regulatory response to important contractual issues was reflected in TRAI's recommendations over time regarding the crucially important spectrum resource. While the ambit of the regulator's recommendatory powers was considerably circumscribed by statute, and was institutionally absent during the first rounds of privatization, it is nevertheless true that TRAI was subdued in its regulatory response to the need for robust economic management of spectrum. It did not carefully interrogate the vexed relation between license and spectrum that had evolved in an ad hoc fashion over the two tenders prior to its creation in 1997 and compounded its meekness by limiting itself to exhorting the licensor to address the problem of inadequate availability of spectrum. The administrative determination of SUC by DoT continued into the second phase between 2004 and 2008 as well, and TRAI had little, if any, role to play in determining either how much spectrum was made available or in how it was to be priced to extract its scarcity value. TRAI also failed to compel executive action on its unified licensing recommendations of January 2005, which were ignored with the abandonment of the Convergence Bill and the change in the political composition of the government. Subsequent developments culminating in the revised UAS licensing guidelines of December 2005 sealed the regulatory and policy compact that had not taken into account the scarcity value of spectrum but had essentially viewed SUC as sufficient payment for granting the licensees usage rights over spectrum.

Perhaps the recommendations made by TRAI in August 2007 were its most controversial. Their primary focus was to recommend that no cap be placed on issuance of UAS licenses. They are also remarkable for, among others, (1) not carrying forward the unified licensing process such that spectrum and license were delinked, (2) continuing with the artificial differentiation between 2G spectrum (in 800, 900, and 1800 MHz bands) and other bands such that administrative pricing of 2G spectrum would persist, and (3) recommending usage of alternative technology by CMTS and UAS licensees using CDMA or

GSM technologies. These recommendations were followed by the rush to hypercompetition with the simultaneous grant of as many as 122 licenses in 2007–8, which became the subject matter of much controversy, many court cases, a CAG report, and Parliamentary investigation given the perceived success of the 3G spectrum auctions that took place later in 2010 in generating unprecedented revenue for the exchequer.

The judgment in the 2G case and the Supreme Court's advisory opinions in the presidential reference that followed are landmark events in the regulatory journey in the telecom sector. This examination of their fallout indicates that the spectrum management regime has taken a turn towards vestment of greater property rights in the resource among licensees or winning bidders since auctions have become mandatory in the medium turn for alienation of spectrum. Whether this is conducive to unlocking the value of this scarce natural resource given the expectations engendered by technological developments that make out a case for the advocacy of the commons approach to spectrum is an open question. While it may be considered academic to ask at this stage of technological progress whether the latest judicial intervention has locked spectrum management into a private-property-rights regime without adequate consideration of the arguments on the side of keeping open a modicum of public access to the resource, it is important to acknowledge the exhibited preference for a property rights paradigm over a commons regime inherent in the policy choice made consequent to the intervention. That choice is likely to complicate the legacy if the case for a commons dispensation is strengthened in the foreseeable future.

Notes

1 This appears to have been done to prevent a cascading effect on the queue for wireless operating license; each UAS licensee's priority for spectrum was to be based on the date of application for allotment of spectrum (see DoT 2011a: A1188), and this staggered signing of UAS licenses was perhaps intended to permit the licensees to apply for spectrum in an orderly fashion.

2 For the record, it must be noted that the Delhi High Court passed an interim order on February 11, 2008, in a writ petition filed by an applicant deemed ineligible for issue of LoI by DoT, making it clear that anything done by DoT and the licensees shall be subject to the final outcome of the case and that DoT shall take into account the seniority of the petitioner and reserve its rights and seniority for grant of license. TDSAT too followed suit for safeguarding the interests of applicants before it. DoT sought the Solicitor General's opinion on February 29, 2007. He opined

on the same day that there is nothing in the courts' orders to preclude signing of UAS licenses as already approved and that since grant of spectrum comes at a later stage, "in relation to circles where availability of [s] pectrum is in doubt, [DoT] would have to consider suitable measures to protect the rights of the petitioners in the event of [their] ultimate success in the petition" (DoT 2011a: A661).

3 DoT had, however, also issued separate guidelines for 3G and BWA services on November 12, 2007. For 3G, these guidelines stipulated that (1) the 2.1 GHz band would be permitted for 3G, (2) 3G spectrum would be granted through "a controlled, simultaneous ascending e-auction, by a specialised agency to ensure transparency in the selection process", (3) besides the initial spectrum charge, recurring SUC of 0.5 per cent/1 per cent of AGR would be payable for the first three years/thereafter, (4) stiff penalties would be levied for non-compliance of rollout obligations, (5) mergers would not be allowed for five years; spectrum trading and reselling would not be permitted, and (6) CDMA spectrum in 800 MHz band for EV-DO (enhanced voice-data optimized) applications would be treated separately from the 2.1 GHz 3G spectrum, and on demand, 2 × 1.25 MHz would be allocated to CDMA operators on paying "an amount proportionate to the highest bid for spectrum in 2.1 GHz band" (TRAI 2008c: 6). In the BWA guidelines, 2.5 GHz band was stipulated for UAS/ISP licensees and BSNL/MTNL, with 2.3 GHz and 3.3–3.4 GHz bands considered for other service providers, with 2 × 10 MHz being the size of each lot. These guidelines required TRAI to revisit its BWA recommendations later since TRAI had recommended allocation of spectrum in the 3.3–3.4 GHz band.

4 The delay has been attributed to politics as well as "to set a high price" (Sridhar 2012: 102).

5 Two private companies, NM Rothschild & Sons (India) Private Limited and DotEcon Limited, were appointed advisors to the government for conduct of the auctions.

6 CAG used the following three criteria: (1) an offer made by STel, then a prospective licensee, to pay a higher price in the shape of additional revenue share for 10 years, an offer it later enhanced; (2) the comparison made by TRAI in recommendations it made in May 2010 between 2G spectrum in the 1800 MHz band and 3G spectrum and suggesting that the 3G auction prices could "be adopted as current price of 2G spectrum in 1800 MHz band"; and (3) the infusion of foreign direct investment in new UAS licensee companies in the form of equity, assuming that the value of a new company with no experience in the telecom sector "can primarily be taken as that of the licence and access to spectrum" (CAG 2010: viii). The linkage with 3G auction prices proved to be the highest estimate.

7 The author was involved in the examination of these recommendations as an official in DoT at the time.

8 In the event, TRAI's expert-determined pan-India price for each MHz of 1800 MHz band spectrum beyond 6.2 MHz turned out to be higher than the 3G auction price, although there were significant differences at the service area level in either direction. While the pan-India per MHz price of 2100 MHz spectrum for a 20-year tenure yielded by the 3G auction was about INR 3,350 crore, the expert-estimated price given in February 2011 for the 1800 MHz band was about INR 4,572 crore for spectrum beyond

6.2 MHz (and about INR 1,770 crore for spectrum up to 6.2 MHz) (TRAI 2011b). The entire exercise establishes the apocryphal nature of administrative determination of spectrum prices. Much effort was later expended by TRAI, during 2013 and 2014, to reset the mechanism for spectrum valuation (see, especially, TRAI 2013b).

9 Indeed, the trial court pronounced its judgment on December 21, 2017, acquitting all accused in the criminal cases. The matter, as can be expected, is under appeal.

10 That the protection of investments was an important factor in making the presidential reference is attested to by a key government functionary at the time, who stated that he convinced the then chief justice of India to take up the reference, arguing that "the country would be faced with humongous claims in international arbitrations under BITs" (Khurshid and Sharma 2018: 138–139, note: BITs refer to bilateral investment treaties). Alas, the court did not answer the specific question related to investments in its opinion.

11 NTP 2012 makes clear the intention to liberalize spectrum "to enable use of spectrum in any band to provide any service in any technology as well as to permit spectrum pooling, sharing and later, trading to enable optimal utilisation of spectrum through appropriate regulatory framework" (NTP 2012: 12).

12 If we assume that DoT – with the help of TRAI – will be able to resolve the host of legacy issues and level the field, the same stronger set of property rights would vest in *all* licensees.

13 Or, more correctly, 'managed commons', following Prasad and Sridhar (2014: 64).

14 Spectrum sharing is often suggested as an attractive transition to a framework of flexible use (Prasad and Sridhar 2014: 142).

15 These include spread spectrum, ultra wideband, opportunistic sharing or agile radio or software defined radio, intelligent coding and smart antennas, and mesh networks (Faulhaber and Farber 2002; Werbach 2004).

16 See for example Demsetz (2002), which takes forward his original approach that relies on markets and private property rights to resolve conflicts over the use of scarce resources. The property rights approach saw communal rights or commons as not being conducive to accurate measurement of costs due to externalities. In his later work, too, Demsetz notes the worldwide trend favouring private ownership over collective ownership due to reduced compactness, improvements in productivity, and increased complexity of resource allocation problems and argues that the private property system "has become more important because its institutional arrangements can cope with the changed nature of economizing problems more effectively than the collective control system" (2002: S665).

17 Heller's (1998) arguments on "anticommons", whereby a resource is prone to underuse because there are too many owners holding rights of exclusion, have been urged with respect to centralized allocation of property rights in spectrum through grant of exclusive licenses. See Werbach (2011).

Part IV

ORGANIZATION AND MICRO-GOVERNANCE OF TRAI

This section deals with the organizational issues emanating from the provisions in the Telecom Regulatory Authority of India (TRAI) Act and thus with the micro-politics of regulatory governance. It moves thereafter to other aspects of institutional design, such as the organizational ethos, the composition of TRAI over the years, the substantive powers vested in TRAI, a timeline of policy and regulatory events, and the culture of participation engendered by TRAI. It discusses the absence of explicit penal powers in the statute and the innovative approach adopted by the regulator to work around this lacuna by using measures such as financial disincentives and other regulatory approaches that have been upheld by the courts.

8

TRAI'S INSTITUTIONAL DESIGN AND ORGANIZATIONAL ETHOS

8.1 Issues in institutional design

The part of the statute governing the design of the telecom regulatory edifice is characteristically over-specified, reflecting the rather maximalist approach to law-making in India. The eight sections from section 3 to section 10 in chapter II of the Telecom Regulatory Authority of India (TRAI) Act, 1997, are intended ostensibly to establish the independence of the economic regulator (as are sections through 14B to 14H and beyond of chapter IV dealing with the Telecom Disputes Settlement and Appellate Tribunal (TDSAT)), yet they have had predictable consequences for matters such as the profile of appointments and terms of office of chairperson and members. What is more analytically salient is whether these legal provisions – and the other formal as well as unstructured rules of the game – have been successful in engendering a culture of independent regulation facilitating thereby ex post maintenance of the institutional arrangements for achieving desired outcomes. What that culture entails in terms of legitimacy and accountability is no doubt shaped by the normative framework of the regulatory enterprise, but it is worthwhile to study the impact of formal and informal rules at the level of the central regulatory organization taking as given the expectations from a set of regulatory arrangements. The way TRAI is structured in turn has important downstream implications for how the 'regulatory game' is played, and this book contends that the nature of institutional design has indeed impacted the regulatory governance of the Indian telecom in the years since TRAI's inception, in terms of both directing the nature of the regulator's interventions on the one hand and eliciting responses from regulated entities and the licensor on the other. What follows is therefore an exploration of the organizational issues surrounding TRAI to better appreciate the structural and behavioural aspects of where and how power inheres in the regulatory system – the 'micro-politics' of governance as it were.

A quick comparison of the statutory provisions before and after the amendment given in Table 8.1 indicates the main aspects of the institutional design pre- and post-amendment. While it is obvious that TRAI lost stature with the creation of TDSAT upon the amendment of the act, such a downgrade was mandated by the courts' view that adjudicatory and non-adjudicatory powers did not sit well within the same institution. A previous chapter noted how the creation of a specialized tribunal was indicative of the embedding of a judicial mentality at the heart of the regulatory architecture as a corrective to legislative overreliance on technical and domain-based expertise as reflected in the statute. This movement of judicial functions to TDSAT aside, the important changes made to the statute included reduction in the number of members and their term of office (and a concomitant reduction in the period of prohibition for taking up commercial employment after demitting office), the introduction of the category of part-time members, the procedure for removal in certain cases, and removing the power of prescribing conditions of service of TRAI's officers and employees from the regulator to the government. In addition, while the provision in section 6 of the TRAI Act for appointing one of the members as vice-chairperson continues unchanged, it is noteworthy that no member has been so appointed subsequent to the amending act, while the pre-amendment TRAI had a functioning vice-chairperson. While these and some other changes appear innocuous at first, they deserve closer examination for a fine-grained institutional analysis of the relationship between the regulator and the government in its twin roles of sovereign and licensor.

8.2 TRAI's organizational ethos

The composition of authority

While institutions encompass the far wider – almost immutable – ambit of norms, conventions, and rules that traverse a field beyond organizations, and the historical context has critical influence over the regulatory decision-making process, it is equally true that organizations are structured and develop procedures that in turn affect the shaping of institutional norms. Organizational structure may not determine how and what decisions are ultimately taken, but it forms, along with institutional norm and individual behaviour, the triptych on which all regulatory decisions are wrought. An organization's response to

Table 8.1 Changes in chapter II of the TRAI Act

Section	*Provision*	*Pre-amendment*	*Post-amendment*
3 (3)	Composition	Chairperson + 2–6 members	Chairperson + 2 whole time members (WTM) + 2 part time members (PTM)
4	Qualification	Chairperson "shall be a person who is or has been" a judge of the Supreme Court (SC) or Chief Justice of a High Court; member "shall be a person who has special knowledge of, and professional experience in" telecom, industry, finance, accountancy, law, management, consumer affairs; provided that "a person who is or has been in the service of Government shall not be appointed as a member unless such a person has held the post of Secretary or Additional Secretary or the post of Additional Secretary and Secretary to the Government of India or any equivalent post in the Central Government or the State Government for a period of three years"	Chairperson and members shall be appointed from among persons who have "special knowledge of, and professional experience in" telecom, industry, finance, accountancy, law, management, consumer affairs; provided that "a person who is or has been in the service of Government shall not be appointed as a member unless such a person has held the post of Secretary or Additional Secretary or the post of Additional Secretary and Secretary to the Government of India or any equivalent post in the Central Government or the State Government for a period of three years" (member includes chairperson under s. 2(f))
5 (2) and (3)	Term of office	Chairperson – 5 years Member – 5 years or until age 65, whichever is earlier	Chairperson and members – 3 years or until age 65, whichever is earlier; incumbents from pre-amendment period to vacate office with three months' compensatory pay
5 (6A)	Allowances of PTM	PTM shall receive such allowances as may be prescribed under s. 35 (2) (aa)	–

(*Continued*)

Table 1.3 (Continued)

Section	*Provision*	*Pre-amendment*	*Post-amendment*
5 (8)	Prohibition on employment	Chairperson and member ineligible for further employment under central or state government and "shall not accept any commercial employment" for a period of 2 years after ceasing to hold office	Chairperson and WTM ineligible for further employment as before for a period of 1 year after ceasing to hold office*; provided incumbents holding office before amendment shall be eligible for re-employment in TRAI or TDSAT
7	Removal	Grounds for removal of member specified; no member shall be removed on the two grounds of (1) having acquired financial or other interest likely to prejudicially affect functions and (2) having abused his or her position so as to render continuance in office prejudicial to public interest unless SC holds enquiry and reports that the member ought to be removed on such grounds; member can be suspended by central government until receipt of SC report (s. 7(2) and (3))	No change in grounds for removal; no member shall be removed from office on the two grounds (no change) unless given a reasonable opportunity of being heard (erstwhile s.7 made applicable to TDSAT as s. 14G)
10 (2)	Terms and conditions of employees' service	The salary and allowances payable to and the other conditions of service of the officers and other employees "shall be such as may be determined by regulations"	The salary and allowances payable to and the other conditions of service of the officers and other employees "shall be such as may be prescribed" as provided in s. 35 (2) (ca); provided pre-amendment regulations shall cease to have effect on notification of rules under s. 35 (2) (ca)

* The section was further amended by ordinance on May 28, 2014, to prohibit employment with government or telecom companies for a period of two years after ceasing to hold office; the explanation at the end was omitted.

Source: Compiled by author.

institutional cues often reflects the desire not merely to correspond with rules (to do the correct thing) but also to achieve internal and external legitimacy for its actions (to do things correctly) so that these actions in turn can constitute a corpus of activity that attains institutional status over the period of the organization's existence. To put it differently, while the law and statute may propose the normative framework for regulatory decision-making, the way in which the regulatory organization is structured and the agency of individual actors within and outside that organization as reflected in their working of the rulebook have a crucial impact on how the institutions of regulation achieve legitimacy in the public eye.

The chapter now turns to an examination of some sections of the existing statute (the TRAI Act) that have bearing on the regulator's micro-governance. Chapter II of the Act begins with section 3, which deals with the establishment and incorporation of the regulator. While clause 3 (3) of the TRAI Bill, 1996, proposed that TRAI shall be composed of a chairperson and between two and four members, the Parliamentary Standing Committee recommended that the strength be increased to a chairperson plus a minimum of four and a maximum of six members to make the composition "more comprehensive and broad based with representation to various interests" (Standing Committee on Communications [SCC] 96–97 1996: 5). Section 3 (3) of the TRAI Act, 1997, as enacted on March 29, 1997, provided for the TRAI to be composed of "a Chairperson, and not less than two, but not exceeding six members" to be appointed by the central government. As indicated in Table 8.2, the amendment Act changed the composition to consist of "a Chairperson, and not more than two whole-time members and not more than two part-time members" to be appointed by the central government.

The pre-amended Act provided for a modicum of structural independence since the chairperson of TRAI was required to be drawn from the judiciary. Given the history of judicial-executive relations in India, it would be difficult to argue that a chairperson who had been a judge of the Supreme Court of India of the chief justice of a High Court would be on the same policy wavelength as the central government, much less in its thrall. The amended Act does not contain this structural feature, and a cursory glance at Table 8.2 reveals the preponderance of ex-officials from government at the helm of the regulator since 2000.[1] In fact, clause 4 (2) of the TRAI Bill, 1996, proposed that members shall be drawn exclusively from senior central or state government bureaucracies, but the final provision in section 4 (2) of the pre-amended Act (carried into section 4 after the amendment) appears

Table 8.2 Chairperson and members of TRAI since inception

	Name	Designation	Age at appointment	Education	Whether ex-government	Service	Tenure		
							From	To	Duration
TRAI 1	Justice S S Sodhi	Chairperson	>62	NA	–	Judge	March 25, 1997	January 24, 2000	2y 10m
	B K Zutshi	Vice-chairperson	59	NA	Yes	Indian Administrative Service (IAS)	March 25, 1997	January 21, 2000	2y 10m
	N S Ramachandran	Member	NA	NA	Yes	Indian Telecom Service (ITS)	March 25, 1997	January 21, 2000	2y 10m
	M Ravindra	Member	58	NA	Yes	Indian Railways Service of Engineers	January 5, 1998	January 21, 2000	2y
	U P Singh	Member	NA	NA	Yes	Indian Revenue Service (IRS)	December 26, 1998	January 21, 2000	1y 1m
	Arun Sinha	Member	NA	NA	Yes	IAS	December 28, 1998	January 21, 2000	1y 1m
	R R N Prasad*	Member	NA	NA	Yes	ITS	January 15, 1999	January 21, 2000	1y
TRAI 2	M S Verma	Chairperson	59	MA, CAIIB	Yes	State Bank of India	March 23, 2000	March 22, 2003	3y
	Ravi Kant	WTM	59	NA	Yes	IRS	March 22, 2000	March 21, 2003	3y
	R R N Prasad*	WTM	NA	NA	Yes	ITS	January 31, 2000	January 31, 2003	3y
	Rakesh Mohan	PTM	52	PhD	NCAER	Economist	April 3, 2000	April 3, 2003	3y
	V S Raju	PTM	NA	Dr. Ing.	IIT Delhi	Engineer	March 27, 2000	March 20, 2003	3y
TRAI 3	Pradip Baijal	Chairperson	60	BE	Yes	IAS	March 22, 2003	March 21,	3y
	D P S Seth	WTM	60	PhD	Yes	ITS	March 10, 2003	March 9, 2006	3y

	P K Sarma	WTM	59	BA, DPF	Yes	IRS	April 10, 2003	April 9, 2006	3y
	Arvind Virmani	PTM	54	PhD	ICRIER	Economist	July 15, 2003	July 14, 2006	3y
	S G Dhande	PTM	55	PhD	IIT Kanpur	Engineer	July 15, 2003	July 14, 2006	3y
TRAI 4	Nripendra Mishra	Chairperson	61	MSc, MA, MBA	Yes	IAS	March 22, 2006	March 21, 2009	3y
	A K Sawhney	WTM	59	BA	Yes	IRS	June 30, 2006	June 29, 2009	3y
	R N Prabhakar	WTM	61	BSc, MBA	Yes	ITS	February 27, 2007	February 26, 2010	3y
	N Balakrishnan	PTM	56	PhD	IISc Bangalore	Engineer	September 26, 2006	–	–
	Rajiv Kumar	PTM	NA	PhD, DPhil	ICRIER	Economist	January 23, 2007	–	–
TRAI 5	J S Sarma	Chairperson	60	PhD	Yes	IAS	May 14, 2009	May 13, 2012	3y
	R Ashok	WTM	60	MSc, MBA	Yes	Indian Railways Accounts Service	September 30, 2009	September 29, 2012	3y
	H S Jamadagni	PTM	NA	PhD	IISc Bangalore	Engineer	March 19, 2010	March 18, 2013	3y
	V Ranganathan	PTM	64	PhD	IIM Bangalore	Economist	May 25, 2011	January 24, 2012	8m
	Pankaj Chandra	PTM	NA	PhD	IIM Bangalore	Management	February 23, 2012	November 25, 2013	1y 9m
TRAI 6	Rahul Khullar	Chairperson	59	PhD	Yes	IAS	May 14, 2012	May 13, 2015	3y
	R K Arnold	WTM	60	BE	Yes	ITS	January 23, 2012	January 22, 2015	3y
	Vijayalakshmy K Gupta	WTM	62	PhD	Yes	Indian Defence Accounts Service	March 4, 2013	NA	NA
	Anil Kaushal	WTM	60	BE, MBA	Yes	ITS	April 7, 2015	NA	NA
	M. Jagadesh Kumar	PTM	NA	PhD	IIT Delhi	Engineer	April 13, 2015	NA	NA

(Continued)

Table 8.2 (Continued)

	Name	*Designation*	*Age at appointment*	*Education*	*Whether ex-government*	*Service*	*Tenure*		
							From	*To*	*Duration*
	Rajiv Sharma	PTM	62	MA	Yes	IAS	April 16, 2015	NA	NA
TRAI 7	R S Sharma	Chairperson	59	MSc, MS	Yes	IAS	August 10, 2015	NA	NA
	Anil Kaushal	WTM	60	BE, MBA	Yes	ITS	April 7, 2015	NA	NA
	H Pradeep Rao	WTM	60	MA, MBA, CIA	Yes	Indian Audit & Accounts Service	September 12, 2017	NA	NA
	M. Jagadesh Kumar	PTM	NA	PhD	IIT Delhi	Engineer	April 13, 2015	NA	NA
	Nitin Kumar Kohli	PTM	NA	MSc, MPhil	Yes	Army	NA	NA	NA
	Abhay Karandikar	PTM	NA	PhD	IIT Kanpur	Engineer	NA	NA	NA

* R R N Prasad was reappointed as whole time member after the TRAI Act was amended (see proviso under section 5 (8)).
Source: TRAI, web resources.

to have been the result of the executive branch's acceptance of the Standing Committee's exhortation that "Chairperson and Members should be drawn from both within and outside Government to ensure transparency, specialised knowledge, experience and understanding of certain disciplines" (SCC 96–97 1996: 5) and the specific recommendation that followed to draw the membership from the fields of telecom, industry, finance, accounting, and consumer affairs with varied experience in public utility services.

The problem with the wording of the erstwhile section 4 (2) and the current section 4 of the TRAI Act extends to two areas: (1) its inclination for seeking out potential appointees from the ranks of government officials and (2) even within that group, its manifest preference for persons who held the position of secretary and/or additional secretary or equivalent to the government of India for at least three years. While the body of section 4 of the amended Act indeed states that the chairperson and members shall be appointed from among persons with special knowledge and professional experience in specified areas of expertise, the proviso that follows maximally delineating the bureaucratic experience required of government officials is a powerful reminder of the executive branch's original intention to seek TRAI's membership from among retiring or retired government officials. Worse, the proviso to section 4 appears to be unique to the telecom sector regulator since – as noted by the Parliamentary Standing Committee in February 2009 – such a proviso "does not find mention in the qualifications for appointment to similar posts in other Regulatory Bodies" (SCIT 08–09 2009: 22) that examined the qualifications prescribed in the case of appointments to other economic regulators such as Securities and Exchange Board of India (SEBI), Insurance Regulatory and Development Authority (IRDAI), Central Electricity Regulatory Commission (CERC), Competition Commission of India (CCI), Petroleum and Natural Gas Regulatory Board (PNGRB), Coastal Aquaculture Authority (CAA), and Forward Markets Commission (FMC) under the respective statutes[2] while examining a proposed amendment to section 4 of the TRAI Act. The statement of objects and reasons of the TRAI (Amendment) Bill, 2008, indicated that "the provisions of the proviso of section 4 have been considered as unreasonable, inconsistent and having no nexus with the object of finding an appropriate person to be appointed as Chairperson or Member of the TRAI as it inflicts a disability in cases where a person who has served the Government, even for short duration, in one or other capacity and while being in Government or after leaving the Government has excelled in the field of telecommunication or other relevant field" only for the reason for the person not having held the post of secretary or additional secretary or equivalent post in government for three years (ibid: 27). The amendment sought to make the condition of three years' service applicable only to superannuating government officials (and not those who may have resigned or quit earlier).

Somewhat typically, at the time of writing, there has been no change in section 4 of the statute since the observations of the Standing Committee were made in February 2009. To be fair, amendment of statute is not to be undertaken lightly, and the Standing Committee's report refers to a number of proposals that have emanated from TRAI for amending portions of the act, and these are perhaps the subject matter of inter-departmental consultations. What is of greater import is the basis written into the normative framework as qualification for heading the economic regulator of the technology-intensive sector. As noted, this is heavily skewed in theory and practice in favour of government officials. While it is an open question whether "the opening up of the telecom sector to the private service providers and the stupendous growth of the telecom services in the [country] are neither a cause nor an effect of the qualifications prescribed for the post of Chairperson/member of TRAI" (ibid: 23), and whether the nature of the Indian state – described in the political economy context of telecom liberalization as "dysfunctional" and "not catalytic" in 1998 (J P Singh 1998: 50), has evolved sufficiently enough to venture private sector participation in leadership roles in economic regulation, there are few good reasons why the zone of consideration for selecting full-time members of TRAI has not expanded beyond the usual suspects from government service. The only instance of a non-government full-time member was immediately after the Act was amended in 2000, when the retired chief of a public sector bank was appointed as chairperson of TRAI.

This appointment brings us face to face with the challenge of finding suitable regulatory personnel at the highest level given the structural factor of a permanent civil service in India that is uniquely situated to manipulate the levers of power within the executive branch. There is some merit in the argument that the appointment of the ex-banker reflected (1) a diminution in the stature of the regulatory authority once a specialized tribunal was put in place and (2) a reduction in the high protocol status the previous incumbent enjoyed since he was a judge. The structure of the Indian bureaucracy becomes a plausible proximate reason for the procession of ex-IAS officers who have been appointed chairperson of TRAI after that first experiment post-amendment.

We turn next to section 5 of the TRAI Act, which underwent a few significant changes upon amendment in 2000. The foremost was a reduction in the terms of office of chairperson and members from the previous five years to three. The term of office corresponded with that prescribed for chairperson and members of the appellate tribunal, although – by virtue of being drawn from the higher judiciary where the age of superannuation is higher than in the civil service – the chairperson of TDSAT was to demit office on attaining the age of 70, while all others (in TRAI and TDSAT) would retire at the age of 65. While the upper age limit for personnel across different regulatory bodies shows a mixture of three-year and five-year tenures,[3] what is of interest in TRAI's case is the reduction in the tenure pre- and post-amendment.[4] Three years is a reasonable duration for as technologically dynamic a sector as telecom, and Table 8.3 indicates the key regulatory decisions taken during the tenure of successive chairpersons in TRAI that can be seen as an inter-temporal comparator of the regulatory responses to key developments in the environment.

A second feature of the amendment relating to conditions of service is the prohibition on further employment under the central or state governments that extended to the chairperson and members in the pre-amendment era but was limited to the chairperson and (whole-time) members of TRAI alone, without being applicable to TDSAT. While section 5 (8) of the TRAI Act is more or less the same as it was prior to the amendment, a corresponding provision was not incorporated for the chairperson and members of TDSAT under the new chapter IV of the act. This had at least one practical fallout: one erstwhile secretary to the government of India came to be appointed as chairperson of TRAI after enjoying one tenure (of about 10 months) as member of TDSAT, thereby bringing into question the perception that TDSAT stood at a higher pedestal as an appellate body. The framework of the TRAI Act that allows for lateral movement of members from TDSAT to TRAI but not the other way around reinforces the perceived importance of the substantive regulatory job content in TRAI over the dispute settlement and appellate nature of the non-judicial members' job in TDSAT. Whether such an arrangement breaches propriety and is conducive to free and fair adjudication of appeals against any direction, decision, or order of TRAI brought before the tribunal is a different matter. It is also a moot question whether the prohibition on appointment should apply equally to all members of TRAI and TDSAT not drawn from the higher judiciary since there exist subtle differences between the provisions of section 4 and section 14C (b) of the act.

Table 8.3 Timeline of policy and key regulatory events

Period	*Chairperson, TRAI*	*Minister(s)*	*Chairmen, TC*	*Key events*
April 25, 1997–January 24, 2000	Justice S S Sodhi	B P Verma (June 1, 1996–March 19, 1998) A B Vajpayee* (March 19, 1998–March 21, 1998) Buta Singh (March 21, 1998–April 19, 1998) A B Vajpayee* (March 19, 1998–April 20, 1998) Sushma Swaraj (April 20, 1998–October 11, 1998) A B Vajpayee* (October 11, 1998–November 6, 1998) Jagmohan (December 6, 1998–June 8, 1999) A B Vajpayee* (June 9, 1999–October 13, 1999) R V Paswan (October 13, 1999–September 1, 2001)	A V Gokak (04.11.96–28.08.98) Anil Kumar (28.08.98–07.02.00)	✓Government announces internet policy (16.09.97) ✓Department of Telecommunications (DoT) gives cellular mobile telephone service (CMTS) license to Mahanagar Telephone Nigam Limited (MTNL) (October 10, 1997) ✓TRAI rules on judicial side that consultation with regulator mandatory (February 18, 1998) ✓Telecommunications Tariff Order (TTO) 1999 issued (March 8, 1999) ✓NTP 1999 comes into effect (April 1, 1999) ✓Telecommunication Interconnection (Charges and Revenue Sharing) Regulation (TIR) 1999 issued (May 28, 1999) ✓Migration package made effective (August 1, 1999) ✓TRAI introduces calling party pays (CPP) regime (November 1, 1999) ✓Amendment ordinance issued (January 24, 2000)

March 23, 2000–March 22, 2003	M S Verma	R V Paswan (October 13, 1999–September 1, 2001) P Mahajan (September 2, 2001–January 28, 2003) A Shourie (January 29, 2003–May 22, 2004)	Anil Kumar (August 28, 1998–February 7, 2000) Shyamal Ghosh (February 7, 2000–May 31, 2002) Vinod Vaish (June 1, 2002–January 31, 2004)	✓ TRAI recommendations on cellular mobile service provider (CMSP) revenue sharing (June 23, 2000) ✓ TRAI recommendations on fixed service provider (FSP) revenue sharing (August 31, 2000) ✓ Bharat Sanchar Nigam Limited (BSNL) formed (October 1, 2000) ✓ TRAI recommends limited mobility (January 8, 2001); DoT issues FSP guidelines (January 25, 2001) ✓ New FSP licenses issued (September 7, 2001) ✓ Fourth CMSP licenses issued (October 9, 2001) ✓ USO Fund established (April 1, 2002)
March 22, 2003–March 21, 2006	Pradeep Baijal	A Shourie (January 29, 2003–May 22, 2004) D Maran (May 23, 2004–May 15, 2007)	Vinod Vaish (June 1, 2002–January 31, 2004) Nripendra Mishra (February 1, 2004–March 31, 2005) Brijesh Kumar (April 1, 2005–June 30, 2005) J S Sarma (July 1, 2005–July 17, 2006)	✓ Subscriber linked criteria (SLC) notified (September 2003) ✓ TRAI recommendations on unified licensing (October 27, 2003) ✓ Cabinet approval for unified access service (UAS)/unified license (UL) (October 31, 2003) ✓ NTP 1999 amended (November 11, 2003) ✓ TRAI clarification on entry fee (November 14, 2003) ✓ DoT decides to accept UASL applications on continuous basis (November 17, 2003)

(Continued)

Table 8.3 (Continued)

Period	*Chairperson, TRAI*	*Minister(s)*	*Chairmen, TC*	*Key events*
				✓TRAI recommendations on unified licensing (January 13, 2005) ✓TRAI recommendations on spectrum-related issues (May 13, 2005) ✓DoT changes national long distance (NLD)/international long distance (ILD) policy (November 2005) ✓DoT issues revised UAS guidelines (December 15, 2005)
March 22, 2006–March 21, 2009	Nripendra Mishra	D Maran (May 23, 2004–May 15, 2007) A Raja (May 16, 2007–November 14, 2010)	J S Sarma (July 1, 2005–July 17, 2006) D S Mathur (July 18, 2006–December 31, 2007) Siddhartha Behura (January 1, 2008–September 30, 2009)	✓SLC revised (March 29, 2006) ✓Group of ministers (GoM) constituted on spectrum (February 23, 2006); Terms of Reference modified (December 7, 2006) ✓TRAI recommendations on third generation (3G)/broadband wireless access (BWA) spectrum (September 27, 2006) ✓DoT reference seeking recommendations on capping UAS licenses (April 17, 2007) ✓TRAI recommendations on not capping licenses (August 28, 2007) ✓DoT announces cut-off dates (September 24, 2007/October 1, 2007) ✓DoT approval for not capping (October 17, 2007)

				✓ Ministry of Finance (MoF)-DoT correspondence (November 2007) ✓ Financial disincentive for unsolicited commercial calls (March 17, 2008) ✓ 3G/BWA guidelines (August 1, 2008)
May 14, 2009–May 13, 2012	J S Sarma	A Raja (May 16, 2007–November 14, 2010) Manmohan Singh* (November 15, 2010) Kapil Sibal (November 16, 10 onwards)	Siddhartha Behura (January 1, 2008–September 30, 2009) P J Thomas (October 1, 2009–September 7, 2010) R Chandrashekhar (September 24, 2010–March 31, 2013)	✓ 3G/ BWA spectrum auctions (April/May 2010) ✓ TRAI recommendations on licensing framework and spectrum management (May 11, 2010) ✓ Comptroller and auditor general (CAG) performance audit report (November 2010)
May 5, 2012–May 13, 2015	Rahul Khullar	Kapil Sibal (November 16, 2010–May 25, 2014) Ravi Shankar Prasad (May 26, 2014 onwards)	R Chandrashekhar (September 24, 2010–March 31, 2013) M F Farooqui (April 1, 2013–June 30, 2014) Rakesh Garg (July 17, 2014– January 28, 2016)	✓ 2G case judgment cancelling licenses (February 2, 2012) ✓ Presidential reference to Supreme Court (April 12, 2012) ✓ NTP 2012 announced (May 31, 2012) ✓ SC opinions in the presidential reference (September 27, 2012) ✓ Recommendations on spectrum valuation and reserve price (September 9, 2013) ✓ 800, 900, 1800, and 2100 MHz spectrum auctions (March 26, 2015)

(*Continued*)

Table 8.3 (Continued)

Period	*Chairperson, TRAI*	*Minister(s)*	*Chairmen, TC*	*Key events*
August 10, 2015onwards	R S Sharma	Ravi Shankar Prasad (May 26, 2014–July 4, 2016) Manoj Sinha (July 5, 2016 onwards)	Rakesh Garg (July 17, 2014–January 28, 2016) J S Deepak (January 21, 2016–February 28, 2017) Aruna Sundararajan (March 1, 2017onwards	✓700, 800, 900, 1800, 2100, 2300, and 2500 MHz spectrum reserve prices ✓Reduction in interconnection usage charges (IUC) ✓Recommendations on spectrum caps ✓Recommendations on net neutrality ✓Inputs for formulation of NTP 2018 ✓Draft National Digital Communications Policy 2018 released (May 1, 2018) ✓Recommendations on data privacy (July 16, 2018)

* Prime minister

Source: Compiled from TRAI and web resources by author.

Apart from the absence of a statutorily laid-down procedure for selection of chairperson and members of TRAI, the withdrawal under section 7 (2) of the previous procedure for removal (for specified reasons) that entailed reference being made to the Supreme Court is another aspect of the amendment attracting critical notice. The position of TRAI in this regard is in contrast to that of TDSAT; section 14B (2) of the Act lays down that chairperson and members of TDSAT shall be appointed in consultation with the chief justice of India, and section 14G dealing with removal of chairperson and members of TDSAT is substantively directly imported from the pre-amendment section 7 requiring reference to the Supreme Court with respect to two of the five specified grounds for removal.[5] While the provision for removal has not been exercised in practice either in TRAI or TDSAT, the difference in procedure – whereby TRAI members get "a reasonable opportunity of being heard" while TDSAT members can be proceeded against only after the Supreme Court reports accordingly after enquiry – is a potential source of executive dominance over the regulator apart from being a dilution of the statutory provision for the independence of the regulator. Once again, the reason for the change in procedure for TRAI can be traced to the events preceding the amendment and the confrontationist position adopted by the pre-amendment TRAI towards DoT even in matters not on the adjudicatory side, including tariff rebalancing and the introduction of CPP on which the regulator took strong positions that militated against the interests of DoT in its role as service provider. It has been pointed out that the pre-amendment Act gave a legally weak mandate to TRAI and at the same time vested it with wide-ranging powers; it would seem that the judicially mandated reform effort intended to focus TRAI's attention on expertise-based regulation while keeping it free of dispute settlement functions (that were hived off to TDSAT) also gave rise to an opportunity for the executive branch to curtail some of the powers and safeguards conducive to fearless exercise of the regulator's non-adjudicatory functions. It is not enough to argue that there is only a remote possibility that these pared down provisions would be exercised by the executive branch – and that they have not indeed been used since 2000. The signal that such derogation of statutory protection and its continuance on the statute book sends to the regulated environment is not conducive to an enhancement of credible commitment to both potential investors and the already invested.

The amendment brought about to the provisions of section 10 of the TRAI Act in 2000 is a sign of either the legislature's excessive confidence in the institution of independent regulation prior to the amendment or the executive's reluctance to countenance *any* deviation from service conditions mandated for government employees. While section 10 (1) of the Act empowering TRAI "to appoint officers and such other employees as it considers necessary for the efficient discharge of its functions under this Act" was retained as it was, section 10 (2), which in the pre-amendment period laid down that the salary and allowances and other conditions of service of the officers and other employees so appointed "shall be as may be determined by regulations", was changed to "shall be as may be prescribed", ensuring that it would be government that would make rules regarding service conditions of TRAI employees unlike previously when TRAI was empowered to do so by regulation. Consequential amendments were also made to the relevant sections by including clause (ca) under subsection (2) of section 35 that gave the central government the power to make rules for carrying out the purposes of the Act and by omitting clause (c) of sub-section (2) of section 36 that gave TRAI the power to make regulations. The amendment appears to have been occasioned by a perception that TRAI had not exercised the provision judiciously in its pre-amendment avatar. A DoT official told the Parliamentary Standing Committee in the year 2000 that "the employees and consultants of the erstwhile TRAI had been engaged with certain perquisites which were far in excess of what was generally considered reasonable in the Government system" and that "the Government felt that no particular organisation should be allowed to create an island of prosperity for itself", leading to the change in provision (SCIT 99–00 2000: 27). The Standing Committee noted that the comptroller and auditor general (CAG) "had taken very serious exception to the same and accordingly, the Government withdrew the powers of TRAI to frame rules and regulations for its own employees" (ibid: 36). While the Standing Committee also said that "in view of the functional character of [TRAI] . . . it would be desirable to allow some flexibility, within permissible limits, as is the case with some other organisations so that TRAI can have the required and timely assistance and guidance of experts/consultants" (ibid), there is no evidence that the rules prescribed by the executive leave much scope for flexibility.

What is of particular interest in the change made to section 10 (2) of the Act is its attribution to audit objections in the pre-amendment period. The original provision became a point of contestation between TRAI and CAG; simultaneously, CAG was also involved in a tussle with TRAI regarding its powers under section 23 of the TRAI Act to audit the regulator's tariff-setting function under section 11 (2). An explanation came to be inserted below section 23 (2) when the

Act was amended which declared that decisions taken by TRAI under section 11 (2) (as well as those taken under sections 11 (1) (b) and section 13), "being matters appealable to the Appellate Tribunal, shall not be subject to audit under this section". This amendment was hotly contested by CAG, which took up its reservations about the proposed amendment with the prime minister as well as with the powerful Public Accounts Committee (PAC) of Parliament, arguing that tariff setting was an executive function for which TRAI needed to be accountable to Parliament, similar to other quasi-judicial determinations such as assessment of taxes and duties. The explanation under section 23 (2) continues on the statute book, however. The firmness displayed by the executive branch to retain this explanation reflected the perceived importance of TRAI's tariff-setting function as a pre-eminent signal to the investing community that it was serious about the business of independent regulation (note that only access services had been opened to private sector participation at this point and competitive long distance service provisioning was still a year or two away) in an environment characterized by humungous incumbency advantages to DoT, which in its role as service provider was sustained by cross-subsidies accruing from overpriced long distance tariffs. More important for the present discussion, the whittling down of TRAI's powers to set the terms and conditions of office of its officers and employees using the same amending occasion represents the executive's balancing act to assuage the auditor's ire. DoT used the amendment as an opportunity to acknowledge CAG's objections regarding salaries and allowances of TRAI's employees so as to retain prescriptive powers with itself on these matters. The licensor, however, chose to ignore CAG's objections regarding the constitutionality of the prohibition on auditing tariff-setting functions on the argument advanced by the then secretary of DoT in 2000 that if "tariff fixation was to be put under audit scrutiny and if the audit went into the question of whether the tariff had been properly fixed or not, conflicts may arise thereby creating a lot of uncertainty" (ibid: 21). This view found resonance in the statement of the newly reconstituted TRAI's first chairperson that audit of tariff setting was a "super regulator kind of approach" and that if it was "left open to yet another authority that could rule or overrule or ask for explanation, then, the very purpose of the existence of a regulator would be defeated" (ibid: 22). That this convergence of views between regulator and licensor did not extend to powers to set service conditions of its employees reflected the state of the power play among TRAI, DoT, and CAG at the time the amendments were proposed and enacted. It is necessary to note that there has been no consistency that such powers might lead to the creation of 'islands of prosperity'; other statutes such as the SEBI Act, 1992 (sections 9 (2) and 30 (2) (b)), and IRDA Act, 1999 (sections 12 (2) and 26 (2) (c)), empower the regulatory body to set out the service terms and conditions of their employees.[6]

The substantive powers

The significant change brought about by the Amendment Act in 2000 was of course to section 11 of the TRAI Act, 1997, and expanding chapter IV, which earlier dealt with the dispute settlement functions of TRAI, to create TDSAT both for adjudicating disputes (among licensor-licensee, two or more service providers, or a service provider and a group of consumers) and to hear appeals against any direction, decision, or order of TRAI. The amendment to section 11 (1) changed the scope and focus of the regulator's activity by creating the categories of recommendatory functions (section 11 (1) (a)) and functions that it "shall discharge" (section 11 (1) (b)) and introducing sections 11 (1) (c) and (d) and five provisos. The changes made to section 11 (1) are a clear indication of the preferences of the executive concerning the substantive content of independent regulation. Under the amended statute, TRAI could make, either *suo motu* or on request from the licensor, non-binding recommendations on eight matters – need and timing of introduction of new service provider; terms and conditions of license; revocation of license for non-compliance; measures to facilitate competition, promote efficiency, and facilitate growth in telecom services; technological improvements in services; type of equipment to be used by service providers; measures for the development of telecom technology and industry; and efficient management of available spectrum. The licensor was mandated by the second proviso to necessarily seek TRAI's recommendations on the first of these two matters "in respect of new licence to be issued to a service provider" subject to a procedure by which the recommendations could be accepted or rejected after one round of reconsideration by the licensor. While not entirely uncontroversial, the logic for including licensing matters under the recommendatory functions of TRAI (and not under section 11 (1) (b) of the act) appears to flow from an understanding that this was a 'sovereign' function. The other recommendatory functions, however, do not meet even this criterion. When queried by the Parliamentary Standing Committee why TRAI's recommendations on these 'non sovereign' functions should not be binding on the government, DoT argued that the interests of competition, efficiency, and technological improvements would be well served by "a two-way traffic, *i.e.* the Department would benefit by TRAI's recommendations and simultaneously would like TRAI to consider the Department's point of view", and that these matters were kept under the recommendatory category in view of the "various aspects of the Indian economy, infrastructure conditions, financial viability of operators etc." (ibid: 7). This argument not only reflects executive reluctance to part with control over a large swathe of regulatory space but also muddled thinking regarding the scope of economic regulation in the first place. It sought to portray a false parity between the licensor and regulator in seeking two-way traffic of points of view between the two while continuing to retain all substantive decision-making powers in the hands of the licensor well beyond those classified as responsibilities of the sovereign. It is time the various aspects cited in support of keeping the non-licensing aspects within recommendatory functions are revisited to examine whether these can be transferred to the regulatory mandate of TRAI in the light of techno-economic developments in the intervening decade and more.

The first of these functions is the ambiguous recommendatory function set out in section 11 (1) (a) (viii) of the TRAI Act related to "efficient management of available spectrum". The retention of considerable executive discretion in the hands of DoT has had deleterious consequences for spectrum valuation, and the moral hazard inherent in giving the regulator non-binding recommendatory powers has been exacerbated in the case of the spectrum resource by decisions taken (and not taken) at different points in the privatization of telecom services, finally culminating in the Supreme Court's judgment in the 2G case that continues to hold the field in spite of the later opinion in a presidential reference due to force of circumstance. The judicial determination in February 2012 had influence in the policy direction since adopted by the government that makes it obligatory to auction spectrum. As frequency spectrum is gradually subjected to market valuation, the extraction of upfront fees (along with usage charges on a revenue share basis) entails a shift from the existing loose and ill-defined quasi-property regime governing spectrum resources to a more robust private property oriented framework giving tangible rights to 'owners' of the bandwidth acquired through public auction.

The judgment in the 2G case ensured that TRAI now has a bigger say in the management of spectrum. It is time the statute settles the issue by giving TRAI an institutional role in spectrum management, either by shifting section 11 (1) (a) (viii) as it is into section 11 (1) (b) or even by other arrangements that would give it a meaningful role in the affairs of the Wireless Planning and Coordination (WPC) wing of DoT. TRAI is best placed to generate discussion on optimal utilization of electromagnetic spectrum going forward so that preparatory work could be initiated for a shift in policy if found necessary at a later date. Spectrum management and the other closely connected technical aspects important for the orderly growth of the sector, such as the three functions assigned to TRAI under its recommendatory role in sections 11 (1) (a) (v), (vi), and (vii) dealing with technological improvements to services, type and inspection of equipment used by service providers, and measures for development of telecom technology and industry respectively, have a large policy-making dimension that may render it impractical to consider placing them entirely under the regulator's control. This retention of the political executive's role in sector development becomes more important with the simultaneous push for regulatory independence that may stoke fears of runaway regulatory action if the final decision-making powers in these areas are vested with the regulator.

A culture of participation

The open consultation process adopted by TRAI has struck root and has even been held out since as a model for other regulatory agencies in the country. This early statement of intent seeks to fulfil the requirement of section 11 (4) of the statute that stipulates that "the Authority shall ensure transparency while exercising its powers and discharging its functions". The focus on transparency and participatory decision-making has abided organizationally and is now composed of (1) allowing all stakeholders and the general public to participate in policy formulation by offering their views whenever sought for, (2) providing for explanatory memorandums in regulations and orders that explain decision bases, and (3) holding regular interactions with registered consumer organizations and non-governmental organizations (NGOs) and conducting seminars and workshops. TRAI also claims wide acclaim for its transparent and participative decision-making procedures.

The agency problem inherent in the non-binding recommendations is accentuated by the *appearance* of transparency and participation as well as of an expertise-infused decision process even as the final decision-making – including the right to pick and choose from or even completely ignore the recommendations – continues to remain in the hands of the executive branch. It is problematic to hope that transparency, participation, and expertise would suffice for independent regulation to flourish in the absence of a culture of mutual trust between the executive and the IRA. The legal framework represented by the TRAI Act falls short of providing incentives for the development of mutual respect and trust by *not* mandating for giving public reasons for the policy decisions taken either on the basis of or in the face of regulatory recommendations.

Notes

1 This was a feature pre-amendment, too, as far as the members were concerned.
2 The years in which the respective acts were promulgated were: SEBI, 1992; IRDA, 1999; CERC, 2003; CCI, 2003; PNGRB, 2006; CAA, 2005; and FMC, 1952.
3 The term of office is five years in CCI, IRDAI, and CERC, for example, and three years in SEBI, FMC, and CAA.
4 Even the tenure of the pre-amendment TRAI practically stood curtailed to about three years since it was disbanded with the promulgation of the ordinance in January 2000.
5 Interestingly, there is no specific statutory provision for relinquishment of office by chairperson and members of TDSAT; there is no equivalent of section 5 (7) that applies to members of TRAI.
6 At the same time, under sections 16 and 17 of the Competition Act, 2003, service conditions of officers and employees assisting CCI shall be prescribed by government; section 91 (3) of the Electricity Act, 2003, requires CERC to specify service conditions with the approval of the central government.

9

PENAL POWERS OF TRAI

9.1 Penal powers

Financial disincentives

Section 12 of the Telecom Regulatory Authority of India (TRAI) Act, 1997, contains TRAI's powers to call for information and explanation from service providers, to make enquiry regarding the affairs of a service provider, and to direct inspection of their books of account or other documents. Section 12 (3) states that every service provider "shall maintain such books of account or other documents as may be prescribed", and section 12 (4) stipulates that TRAI "shall have the power to issue such directions to service providers as it may consider necessary for proper functioning by service providers". Section 13 of the Act provides that TRAI may issue directions to service providers from time to time, "as it may consider necessary", with regard to TRAI's non-recommendatory functions.[1] These provisions are supplemented by section 29 of the act, which states that a person violating TRAI's directions "shall be punishable with fine" as specified therein. Section 34 (1) prohibits cognizance of "any offence punishable under this Act or the rules or regulations made thereunder save on a complaint made by [TRAI]"; section 34 (2) stipulates that "no court inferior to that of a Chief Metropolitan Magistrate or a Chief Judicial Magistrate of first class shall try any offence punishable under this Act".

Taken together, these statutory provisions require TRAI to initiate a formal legal process for ensuring that regulated entities comply with its directions.[2] The TRAI Act thus does not directly vest enforcement powers in the regulator; it needs to approach the appropriate civil court if its directions are violated. This could be long drawn out, and the attendant costs make the arrangement a poor substitute for effective enforcement powers vested in the regulator. The absence of enforcement provisions in the statute is a legacy of the pre-amendment act; in fact, after the amendment in 2000, the Parliamentary Standing Committee enquired with the secretary of the Department of Telecommunications (DoT) "the reasons for not giving [TRAI] any power to enforce its decisions" (Standing Committee on Information Technology [SCIT] 99–00 2000: 15). The committee concluded that "although the TRAI has been vested with important regulatory powers, in reality it would feel handicapped in taking the defaulters to task", and recommended that the government should consider whether "it would not be desirable to empower TRAI itself to enforce compliance of its orders, if necessary, by imposing penalty on the defaulters" (ibid: 32). The under-delegation of regulatory power by statute in this case even after the amendment made provision for establishing the Telecom Disputes Settlement and Appellate Tribunal (TDSAT) to theoretically oversee the exercise of such powers on the judicial side is a peculiar feature of the regulatory structure in Indian telecom. This is in contrast with the statutory powers vested in financial regulators such as Reserve Bank of India (RBI) and Securities and Exchange Board of India (SEBI)[3] enabling them to levy penalties on regulated entities for regulatory violations.

Involved as it is in a number of activities mandated under section 11 (1) (b) of the TRAI Act, the regulatory authority can be expected to face violations of directions and stipulations laid down by regulations. The option of approaching the courts is not only time-consuming, but it is also ill-suited to the telecom sector milieu in view of (1) the long-term incomplete nature of contracts where adopting adversarial positions in litigation steeply increases the costs of introducing adaptations to changing circumstances, (2) the technological uncertainties and other information asymmetries requiring the regulator to work in close cooperation with regulated entities, and (3) the procedural uncertainties associated with common law courts. It is trite that regulators require a modicum of enforcement powers to ensure that the regulatory agenda is implemented at least cost; in the absence of explicit statutory provisions for enforcement, such powers are often acquired in an indirect manner.

TRAI has adopted an innovative legal stratagem to assert its powers to impose monetary fines on regulated entities for violation of its directions and regulations. The range of regulated activities that could entail penalties includes violation of regulations regarding unsolicited commercial communication, tariff orders, quality of service (QoS) regulations, regulations on accounting separation, and mobile number portability (MNP) regulations. To overcome the absence of powers to impose penalties directly, TRAI has used what it refers to as 'financial disincentives' that are imposed on regulated entities. TRAI's intent to assert its enforcement jurisdiction for violation of its orders was first manifested in the Telecom Unsolicited Commercial Communications (Amendment) Regulations, 2008, issued on March 17, 2008. These were followed by the simultaneous issuance of the 52nd amendment to the Telecommunications Tariff Order (TTO) (TRAI 2012e) and the fourth amendment to its MNP regulation (TRAI 2012f) on September 19, 2012, after a hiatus of over four years in adopting that route.[4] In both these notifications, the consequences of delay in complying with or contravention of provisions of regulations came to be specified in monetary terms, "by way of financial disincentive . . . that [TRAI] may by way of order direct". A similar dispensation specifying financial disincentives to be levied on service providers for failure to submit reports or statements (or for furnishing false statements) and for failure to meet QoS benchmarks came to be notified on October 15, 2012, and November 8, 2012, as the (first) amendment to the accounting separation regulation (ASR) (TRAI 2012g) and the second amendment to the QoS regulations (TRAI 2012h), respectively.

The introduction of financial disincentives into TRAI's regulatory arsenal in 2008 arose from a realization that the principal regulations were being flouted by regulated entities and that consumers continued to be harassed and subjected to the nuisance of unsolicited commercial communication from telemarketers. TRAI revisited its position articulated in the principal regulations that it had neither the power to impose penalty on defaulters nor the statutory authority to adjudicate disputes in the matter. It stated in the explanatory memorandum (EM) to the amendment regulations in March 2008 that it had been advised that "[TRAI] has sufficient powers, under sub-clause (i) and sub-clause (v) of clause (b) of sub-section (1) of section 11, read with section 13 and section 36 of the [TRAI Act] to ensure the quality of service provided by the service providers and protect the interests of the consumers" (TRAI 2008h: 6). The memorandum also stated that judicial pronouncements[5] indicated that "regulation of a particular matter embraces within its fold the power to do all those things necessary and incidental for regulating that matter" and that unless some form of financial disincentive was levied, "it would not be possible to ensure the quality of service so as to protect the interests of the consumers of the telecom services" (ibid). The explanatory memorandum also noted that any person aggrieved by TRAI's order for payment of financial disincentive could appeal to TDSAT under section 14A of the TRAI Act.

TRAI has been successful in imposing financial disincentives on regulated entities that fall foul of its regulations and tariff orders. The proposition that the TRAI Act vests powers in TRAI to impose such disincentives has been tested and upheld before the Delhi High Court in a case pertaining to regulation of call drops (Tangirala 2016). While the case law adduced by TRAI in support of its regulatory enforcement technique may have logical appeal, it cannot be a substitute for explicit statutory provisions empowering it to levy penalties under section 29 for violation by regulated entities of stipulations laid down by TRAI pursuant to discharging its functions under section 11 (1) (b) and section 11 (2) of the TRAI Act.

Directions and regulation-making powers

The discussion on the ambit of section 29 of the TRAI Act that is at the centre of TRAI's regulatory enforcement framework would be incomplete without adverting to two judgments of the Supreme Court that have a bearing on the regulator's ability to levy financial disincentives for infractions committed by service providers. The first of these is the case *Telecom Regulatory Authority of India* v. *BPL Mobile Cellular Ltd. and Ors.*, Civil Appeal no. 6743 of 2003, in which judgment was delivered by a three-judge bench of the apex court on March 28, 2006; the second is once again a judgment of a three-judge bench in *Bharat Sanchar Nigam Ltd.* v. *Telecom Regulatory Authority of India and Ors.*, Civil Appeal no. 5253 of 2010, delivered on December 6, 2013.

TRAI v. *BPL Mobile* related to an instance of violation by an access service provider of the provisions of TTO in October 2001 that was taken to court by a subscriber as public interest litigation.[6] Pursuant to court directions, TRAI ruled in an order dated December 23, 2002, that the violation by the service provider was of a serious nature and had adversely affected consumer interest. TRAI proceeded to lodge a complaint under section 29 and section 34 on February 4, 2003, before the chief metropolitan magistrate, Bangalore, thereby initiating penal action as stipulated by statute. The service provider challenged TRAI's order of December 2002 before TDSAT; the tribunal upheld the appeal on May 13, 2003,[7] and stated that it was "not convinced that there was any statutory direction which was violated by [the service provider]". TDSAT found that "even though there has been a violation of the reporting requirement under [TTO 1999], the action taken by [TRAI] in filing a complaint before the [magistrate] is not as per law". TDSAT quashed and set aside the order dated December 23, 2002, and directed TRAI to withdraw its complaint before the magistrate. TRAI challenged this order and the interpretation of the language and provisions of section 29 given by TDSAT before the Supreme Court.

In its order disposing of the appeal on March 28, 2006, the three-judge bench of the Supreme Court (Justices Ruma Pal, B N Srikrishna, and Dalveer Bhandari) dealt with the question of whether the word 'directions' appearing in section 29 included TTO such that "any violation thereof would be punishable under Section 29 read with Section 34". The court noted that the TRAI Act uses the terms 'order', 'regulation', and 'directions' in different provisions such as section 11 (1) (c), section 11 (2), and section 12 (4). To answer the question, the order referred to the wording of section 13 of the Act that empowers TRAI to issue directions from time to time to service providers for discharge of its functions under section 11 (1) (b). The court thereafter interpreted TTO 1999 as an order issued to ensure compliance by service providers of the terms and conditions of licence (i.e., section 11 (1) (b) (i) of the act), since the licence prescribed that the licensee would charge tariffs according to TRAI's tariff orders and would also abide by publication requirements and so forth set by TRAI under the act. The court ruled that on a reading of the relevant provisions of the TRAI Act, it found that "the word 'directions' had been used in a wide sense to cover orders/regulations which in effect direct an action to be taken" and that if section 29 was limited "only to directions which were not directory orders or directory regulations this would mean that violation of such orders/regulations would not carry any penal consequence whatsoever". It noted that this would mean that "the entire scheme of the Act would become unworkable"; the order also noted that section 11 (1) (b) has been "widely framed" and the principle that section 29 must be strictly construed since it creates an offence does not militate against interpreting the term 'directions' contextually. The order held that the term "can take within its fold directory orders and regulations in the nature of directions as a matter of semantics" and that there was no reason not to include orders and regulations under its ambit. It respectfully cited the language used by a constitution bench in *Bashiruddin Ashraf* v. *The Bihar Subai Sunni Majlis-e- Awqaf and Ors.*, AIR 1965 SC 1206 that the words 'orders' and 'directions' only differ in degree and that "an order is more peremptory than a direction and an argument can never be right which suggests that while disobedience of a direction should merit the punishment . . . disobedience of an order should go unpunished". The Supreme Court consequently allowed TRAI's appeal.[8]

The second judgment that has a bearing on the ambit of TRAI's regulation-making power, dated December 6, 2013, of the Supreme Court (in *BSNL* v. *TRAI and Ors*., Civil Appeal no. 5253 of 2010), arose from a bunch of appeals[9] filed before it regarding different actions of TRAI consequential to various regulations made by the regulator and some directions given to service providers for ensuring compliance.[10] While the three-judge bench (Justices G S Singhvi, B S Chauhan, and Fakkir Mohamed Ibrahim Kalifullah) took up the matter in view of a reference from a smaller bench regarding five substantial questions of law of public importance, it was agreed that a preliminary issue related to the jurisdiction of TDSAT could be decided before the questions referred to it by the two-judge bench were taken up. The court accordingly framed and answered the following question, which is of considerable importance for the present discussion regarding the powers enjoyed by TRAI:

> Whether in exercise of the power vested in it under section 14 (b) of the [TRAI] Act, TDSAT has the jurisdiction to entertain challenge to the regulations framed by [TRAI] under section 36 of the Act.

The court's order observed that there is nothing in the language of section 36 (2) restricting the scope of section 36 (1) and that "it is settled law that if power is conferred upon an authority/body to make subordinate legislation in general terms, the particularization of topics is merely illustrative and does not limit the scope of general power". It held that sections 12 (4) and 13 of the TRAI Act "are only enabling provisions" and parsed the regulation-making power given to TRAI under section 36 (1) with the powers to issue directions to service providers given under those sections. It stated that TRAI can make regulations under section 36 (1) "empowering it to *issue directions of general character applicable to service providers and others* and it cannot be said that by making regulations under Section 36 (1) [TRAI] has encroached upon the field occupied by Sections 12 (4) and 13 of the Act" (emphasis added). The court observed that the regulation-making power under section 36 is non-delegable under section 33 of the Act since "the power under Section 36 is legislative as opposed to administrative"; by virtue of section 37 that requires them to be laid before Parliament, the regulations made under section 36 "are placed on par with the rules framed by the Central Government under Section 35" and are in the nature of subordinate legislation. The court accordingly held that the power vested in TRAI under section 36 (1) "is wide and pervasive" and "not controlled or limited by Section 36 (2) or Sections 11, 12 and 13" of the TRAI Act.

The court's determination that regulations made under section 36 of the Act are in the nature of subordinate legislation on par with the rules made by the central government under section 35 is crucial to the appreciation of its answer to the question of whether TDSAT had jurisdiction over challenges to TRAI's regulations. As subordinate legislation, regulations issued under section 36 were immune to appeal under the appellate jurisdiction of TDSAT under section 14 (b), which was reserved for administrative and quasi-judicial directions, decisions, or orders; TRAI's legislative actions could not be challenged except in the exercise of judicial review. On a comparative examination of the provisions of the pre-amendment TRAI Act and the amended version, the court held that (1) the original jurisdiction of TRAI was vested in TDSAT by virtue of section 14 (b) and (2) the appellate jurisdiction that was vested in the High Court prior to amendment was given to TDSAT in section 14 (b) and that since the "High Court while hearing appeal did not have the power of judicial review of subordinate legislation, the transferee adjudicatory forum, i.e., TDSAT cannot exercise that power under section 14 (b)". While doing so, it distinguished the previous Supreme Court judgment in *TRAI* v. *BPL Mobile* by observing that the bench which decided that matter "felt that the view taken by TDSAT would encourage rampant violation of orders without any penal consequence and the entire scheme of the Act would become unworkable". It went on to state that, in the earlier judgment,

> The word "directions" used in Section 29 of the Act was interpreted to include orders and regulations in the context of the factual matrix of that case and the apprehension of the Court that Section 29 would otherwise become unworkable, but the same cannot be read as laying down a proposition of law that the words "direction", "decision" or "order" used in Section 14 (b) would include regulation[s] framed under Section 36, which are in the nature of subordinate legislation.

The court proceeded to answer the question it had formulated in the negative, holding that TDSAT did not have the jurisdiction to entertain challenges to regulations made under section 36 of the TRAI Act. At the same time, it was clarified that the regulations could be challenged by aggrieved persons "by filing appropriate petition before the High Court".

On bare perusal, the judgments in *TRAI* v. *BPL Mobile* and *BSNL* v. *TRAI* indicate that (1) as far as section 29 dealing with the penalty for contravention of TRAI's directions is concerned, the term 'directions' is wide enough to cover orders and regulations; and (2) for the purpose of section 14 (b) related to the appellate jurisdiction of TDSAT, the use of the terms 'direction', 'decision', or 'order' does not cover regulations issued under section 36 of the Act since such regulations are in the nature of subordinate regulation that cannot be adjudicated upon by TDSAT. The court has protected TRAI's capacity to penalize violations of its mandates in the first case and has placed restrictions on TDSAT's ability to question the regulations themselves in the second.

The protection available to TRAI to impose penalties, however, is of limited utility since the TRAI Act provides for a cumbersome procedure under section 34 that involves filing a complaint before a designated magistrate. This represents a palpable omission in the design of the statute and could be rectified by amending the Act to provide statutory backing for TRAI to levy penalties by itself, subject to procedural safeguards as provided, for example, in the case of SEBI. In the absence of such explicit power, TRAI has resorted to the mechanism of imposing financial disincentives specified by regulations made under section 36 of the act. The judgment in *BSNL* v. *TRAI* offers a measure of protection against TDSAT questioning or reviewing these (and other) regulations; all the implementation of the provisions for financial disincentives requires to survive appellate challenge before the tribunal is that they follow the attributes expected of administrative action such as natural justice and absence of caprice. The sword of judicial review of the regulations still remains sheathed and unused; aggrieved parties could challenge the *vires* of the regulations imposing these disincentives before the constitutional courts by way of judicial review. It would be speculative to hazard a guess regarding the outcome of such a challenge since it would certainly depend upon the specific factual matrix of the case. Under the circumstances, it appears that an amendment of the Act to provide for TRAI to directly impose penalties on parties violating its orders, directions, decisions, or regulations (suitably hedged with procedural safeguards) would be the best solution to enhance regulatory credibility.

9.2 Micro-aspects: summary and conclusion

The relevance of a micro-level examination of TRAI's organizational structure is important not only to understand its impact on regulatory decision-making but also to impart granularity to the institutional and legal analysis of the regulatory governance framework. We have also noted at the outset that such an examination is useful for a fuller understanding of the role of the regulator in the relations of exchange and interdependence in the regulatory environment. A study of the structural features can also be expected to yield insight into regulatory legitimacy and accountability, two key aspects of the culture of independent regulation. In light of the recent initiatives such as the draft Regulatory Reform Bill, 2013, to bring uniformity into the governing framework of economic regulation across sectors, closer scrutiny of the statutory provisions pertaining to TRAI and its organizational features presents an opportunity to test the essential features of the 'new' regulatory philosophy against what obtains in practice.

Statutory provisions such as those in chapter II of the TRAI Act that govern the regulator's structure, eligibility, terms and conditions of its members and staff, and so forth carry the imprint of the inorganic amendment of the Act in 2000 in response to the courts' rejection of vesting adjudicatory functions in the same expert body tasked with economic regulation previously. Without prejudice to the utility of separating the adjudicatory jurisdiction from the regulatory functions (including recommendatory functions), the creation of TDSAT and the consequential changes to the Act gave rise to certain statutory shortfalls, silences, and gaps in the structure of TRAI's organization. While TRAI is no doubt a more compact body post-amendment, the striking dependence exhibited in practice on ex-government officials – and those of a particular extraction and vintage – to fill the available whole-time positions in TRAI (and for that matter in the secretariat) is one example of the structural shortcoming that survives in the statute, observations of a Parliamentary committee notwithstanding. This penchant for employing regulatory personnel of a particular kind can be expected to have an impact on the nature and quality of regulatory decision-making. The emphasis on candidates' age and experience has the effect of depleting the zone of consideration against the interests of younger aspirants – even more in the case of those who are not government employees – especially since lower bounds and minimum experience criteria have not been stipulated.

The amendment to the Act in 2000 retained some amount of anomaly in the terms and conditions of employment of members of TRAI and TDSAT. While the tribunal is statutorily and theoretically an appellate forum situated at a higher level than the regulator, the peculiar condition in the Act that permits migration of members of TDSAT (but not its chairperson, who is naturally constrained in this regard by being a judicial officer) to TRAI but not in the reverse saw one recent instance of an erstwhile member of the appellate tribunal assuming the chair of the regulatory authority. The absence of a statutory mechanism for selection of TRAI members in contrast to TDSAT is exacerbated by the relatively easier removal process that encourages executive dominance over the regulatory body. These statutory infirmities combined with the ostensibly wide-ranging regulatory mandate given to TRAI illustrate the potential for damage that executive waywardness can wreak on regulatory certainty and credible commitment in the sector.

Another instance of arrogation by the executive branch of what otherwise is expected to be the exercise of legitimate authority by TRAI is the amendment to section 10 (2) of the Act whereby the terms and conditions of officers and staff appointed in the regulatory body are determined by the central government. In the context of the scuffle over excluding audit of TRAI's tariff functions that was also effected by the amending act, the removal of the regulator's power of determining its employees' terms reflected the politics of TRAI-DoT-CAG relations at the time; the retention of such a provision in the TRAI Act cannot even be termed standard practice since SEBI and IRDAI can set out similar service terms and conditions. The tilting of the balance of administrative power in favour of the line department over the independent regulatory authority (IRA) occurred during the same process by which clarity was imparted to different parts of the fledgling regulatory structure under the auspices of court decisions. This feature marked DoT-TRAI relations over the next 15 years and had important effects on the way the regulated entities have perceived their roles inter se.

The discussion on the judgments of the Supreme Court in *TRAI* v. *BPL* and *BSNL* v. *TRAI* serves to illustrate the gaps in the TRAI Act regarding the enforcement powers of TRAI for violations of its regulatory directions. While these and the line of court cases cited in TRAI's regulations appear to offer support to the imposition of financial disincentives for violations of directions, orders, and regulations by regulated entities, the sword of judicial review still hovers ominously over the exercise of this power by the regulator. The statute's insistence on recourse to the general judicial process for penalizing violations of regulatory orders tends to militate against the independence of the regulatory dispensation; what cannot be effectively enforced can hardly be regulated properly. Such an arrangement also weakens the triadic dispute resolution (TDR) ability of the regulator that is – quite apart from and beyond the statutory provisions for adjudication of disputes between regulated entities – at its very institutional core.

Notes

1 A proviso was inserted under section 13 by the Amendment Act of 2000 that the directions under sections 12 (4) and 13 shall be limited to matters specified in section 11 (1) (b) of the amended act.

2 In addition, the language of section 34 (1) indicates that TRAI could file a complaint before the appropriate court requesting cognizance of an offence that is punishable under any rule or regulation made under the TRAI Act.

3 Chapter VIA was introduced in the SEBI Act, 1992, by amendment with effect from January 25, 1995; it contains provisions for levying penalties in the event of defaults, insider trading, etc., and for the appointment of an adjudicating officer with powers to summon and enforce attendance, etc. Orders made by the adjudicating officer are appealable to a Securities Appellate Tribunal established under the provisions of chapter VIB of the act.

4 Of these four, the 52nd amendment order was, however, the first to come into force since it was applicable from the date of publication; the fourth amendment to the MNP regulation took effect after 60 days.

5 The explanatory memorandum referred to *UP Cooperative CU Federations* v. *UP Sugar Mills Association* (AIR (2004) SC 3697).

6 *Sudhanshu* v. *Union of India and Ors.*, Writ Petition no. 656 of 2002 before the High Court of Judicature of Bombay, Bench at Aurangabad. The High Court directed TRAI on September 24, 2002, to take appropriate steps in accordance with law after hearing both parties.

7 *Appeal no. 2 of 2003.*

8 The proceedings before the magistrate's court had already been withdrawn by TRAI in the absence of a stay order on TDSAT's judgment of May 2003. The Supreme Court order made it clear that there would be no revival of those proceedings.

9 The judgment deals with 36 civil appeals and one transfer case.

10 The genesis of some of the appeals can be traced to the Delhi High Court judgment of January 17, 2000, in *MTNL* v. *TRAI*, 2000 Indlaw Del 330 that was discussed in chapter 3.

BIBLIOGRAPHY

Anant, T C A and Jaivir Singh (2007): "Structuring Regulation: Constitutional and Legal Frame in India", in Jaivir Singh (ed.), *Regulation, Institutions and the Law* (New Delhi: Social Science Press), pp. 10–27

Athreya, M B (1996): "India's Telecommunications Policy: A Paradigm Shift", *Telecommunications Policy*, Vol. 20, No. 1 (January–February), pp. 11–22

Ayers, Ian and John Braithwaite (1995): *Responsive Regulation: Transcending the Deregulation Debate* (Oxford: Oxford University Press)

Baldwin, Robert, Colin Scott, and Christopher Hood (1998): "Introduction", in Baldwin, Robert, Colin Scott, and Christopher Hood (eds.), *A Reader on Regulation* (Oxford: Oxford University Press)

______, Martin Cave, and Martin Lodge (2010): "Introduction: Regulation – the Field and the Developing Agenda", in Baldwin, Robert, Martin Cave, and Martin Lodge (eds.), *The Oxford Handbook of Regulation* (Oxford: Oxford University Press), pp. 3–16

______ (2012): *Understanding Regulation: Theory, Strategy, and Practice*, Second Edition (Oxford: Oxford University Press)

Bator, Francis M (1958): "The Anatomy of Market Failure", *Quarterly Journal of Economics*, Vol. 72, No. 3, pp. 351–379

Becker, Gary (1983): "A Theory of Competition among Pressure Groups for Political Influence", *Quarterly Journal of Economics*, Vol. 98, No. 3 (August), pp. 371–400

Black, Julia (1997): "New Institutionalism and Naturalism in Socio-Legal Analysis: Institutionalist Approaches to Regulatory Decision Making", *Law & Policy*, Vol. 19, No. 1 (January), pp. 51–93

——— (2002a): "Critical Reflections on Regulation", *Australian Journal of Legal Philosophy*, Vol. 27, pp. 1–35

——— (2002b): "Regulatory Conversations", *Journal of Law and Society*, Vol. 29, No. 1 (March), pp. 163–196

——— (2008): "Constructing and Contesting Legitimacy and Accountability in Polycentric Regulatory Regimes", *Regulation & Governance*, Vol. 2, pp. 137–164

Braithwaite, John (2006): "Responsive Regulation and Developing Economies", *World Development*, Vol. 34, No. 5, pp. 884–898

Breyer S (1998 [1982]): "Typical Justifications for Regulation", in Robert Baldwin, Colin Scott, and Christopher Hood (eds.), *A Reader on Regulation* (Oxford: Oxford University Press), pp. 59–92

Bruff, Harold H (1991): "Specialized Courts in Administrative Law", *Administrative Law Review*, Vol. 43, No. 3, pp. 329–366

Calabresi, Guido and A Douglas Melamed (1972): "Property Rules, Liability Rules, and Inalienability: One View of the Cathedral", *Harvard Law Review*, Vol. 85, No. 6 (April), pp. 1089–1128

Chakraborty, Chandana and Banani Nandi (2011): "'Mainline' Telecommunications Infrastructure, Levels of Development and Economic Growth: Evidence from a Panel of Developing Countries", *Telecommunications Policy*, Vol. 35, No. 5 (June), pp. 441–449

Chowdary, T H (1998): "Politics and Economics of Telecom Liberalization in India", *Telecommunications Policy*, Vol. 22, No. 1 (February), pp. 9–22

Coase, R H (1960): "The Problem of Social Cost", *Journal of Law and Economics*, Vol. 3 (October), pp. 1–44

Codding Jr., George A (1995): "The International Telecommunications Union: 130 Years of Telecommunications Regulation", *Denver Journal of International Law and Policy*, Vol. 23, No. 3, pp. 501–511

Comptroller and Auditor General of India (CAG) (2000): *Report No. 6 of 2000 for the Period Ended March 1999 Post and Telecommunications* [Available at saiindia.gov.in]

——— (2010): *Performance Audit Report on the Issue of Licences and Allocation of 2G Spectrum by the Department of Telecommunications, No. 19 of 2010–11* (New Delhi: Comptroller and Auditor General of India)

Cooper, Robert M (1938): "Administrative Justice and the Role of Discretion", *Yale Law Journal*, Vol. 47, No. 4 (February), pp. 577–602

Cooter, Robert and Thomas Ulen (2012): *Law & Economics*, Sixth Edition (Boston, MA: Addison-Wesley)

Cramton, Roger C (1964): "The Effectiveness of Economic Regulation: A Legal View", *American Economic Review*, Vol. 54, No. 3 (May), pp. 182–191

——— (1972): "A Comment on Trial-Type Hearings in Nuclear Power Plant Siting", *Virginia Law Review*, Vol. 58, No. 4 (April), pp. 585–599

Cronin, F J, E B Parker, E K Colleran, and M K Gold (1991): "Telecommunications Infrastructure and Economic Growth: An Analysis of Causality", *Telecommunications Policy*, Vol. 15, No. 6 (December), pp. 529–535

Demsetz, Harold (1967): "Toward a Theory of Property Rights", *American Economic Review*, Vol. 57, No. 2 (May), pp. 347–359

——— (1968): "Why Regulate Utilities?" *Journal of Law and Economics*, Vol. 11, No. 1 (April), pp. 55–65

——— (2002): "Toward a Theory of Property Rights II: The Competition between Private and Collective Ownership", *Journal of Legal Studies*, Vol. 31, No. S2 (June), pp. S653–S672 den Hertog, Johan (1999): "General Theories of Regulation", in Bouedewijn Bouckaert and Gerrit de Geest (eds.), *Encyclopedia of Law and Economics* (Cheltenham:University of Ghent and Edward Elgar), pp. 223–270

Department of Telecommunication, Government of India (DoT) (2009a): *Report of the Committee for 'Allocation of Access (GSM/CDMA) Spectrum and Pricing'*, May 2009 (New Delhi: Government of India, DoT)

——— (2009b): *Auction of 3G and BWA Spectrum: Revised Information Memorandum*, October 23, 2009 [Available at www.dot.gov.in]

——— (2010a): *Auction of 3G and BWA Spectrum: Notice Inviting Applications*, February 25, 2010 [Available at www.dot.gov.in]

——— (2011a): *Report on Examination of Appropriateness of Procedures Followed by Department of Telecommunications in Issuance of Licences and Allocation of Spectrum During the Period 2001–2009*, January 31, 2011, pp. 1–143, A1-A1309 (New Delhi: Department of Telecom)

Dokeniya, Anupama (1999): "Re-forming the State: Telecom Liberalization in India", *Telecommunications Policy*, Vol. 23, No. 2 (March), pp. 105–128

Dossani, Rafiq (2002): "Telecommunications Reform in India", *India Review*, Vol. 1, No. 2, pp. 61–90

Dua, Pami and Rajiv Ranjan (2010): "Exchange Rate Policy and Modelling in India", *Study No. 33* (Mumbai: Development Research Group, Department of Economic Analysis and Policy, Reserve Bank of India)

Faulhaber, Gerald R and David J Farber (2002): "Spectrum Management: Property Rights, Markets, and the Commons", *AEI-Brookings Joint Center for Regulatory Studies Working Paper No. 02–12*, December

Fenster, Mark (2006): "The Opacity of Transparency", *Iowa Law Review*, Vol. 91, No. 3 (March), pp. 885–949

Ferejohn, John (2002): "Judicializing Politics, Politicizing Law", *Law and Contemporary Problems*, Vol. 65, No. 3 (Summer), pp. 41–68

Fuller, Lon L (1978): "The Forms and Limits of Adjudication", *Harvard Law Review*, Vol. 92, No. 2 (December), pp. 353–409

Ginsburg, Tom (2008): "Judicialization of Administrative Governance: Causes, Consequences and Limits", *National Taiwan University Law Review*, Vol. 3, No. 2 (September), pp. 1–29

Goldberg, Victor P (1976): "Regulation and Administered Contracts", *Bell Journal of Economics*, Vol. 7, No. 2 (Autumn), pp. 426–448

Group on Telecom and IT Convergence (GoT-IT) (2001): *Report of the Group on Telecom and IT Convergence on Limited Mobility* [Available at pib.nic.in/focus/foyr2001/foapr2001/rg_tit.pdf]

Guasch, J Luis and Pablo Spiller (1999): "New Frontiers in Regulation in the Communications Sector: Managing and Regulating the Spectrum", in *Managing the Regulatory Process: Design, Concepts, Issues, and the Latin America and Caribbean Story* (Washington, DC: World Bank), pp. 225–236

Gupta, Rajni (2002): "Telecommunications Liberalisation: Critical Role of Legal and Regulatory Regime", *Economic and Political Weekly*, Vol. 37, No. 17 (April–May), pp. 1668–1675

Hancher, L and M Moran (1998 [1989]): "Organizing Regulatory Space", in Robert Baldwin, Colin Scott, and Christopher Hood (eds.), *A Reader on Regulation* (Oxford: Oxford University Press), pp. 148–172

Hardy, Andrew P (1980): "The Role of the Telephone in Economic Development", *Telecommunications Policy*, Vol. 4, No. 4 (December), pp. 278–286

Hazlewood, Arthur (1953): "The Origin of the State Telephone Service in Britain", *Oxford Economic Papers*, Vol. 5, No. 1 (March), pp. 13–25

Heller, Michael A (1998): "The Tragedy of the Anticommons: Property in the Transition from Marx to Markets", *Harvard Law Review*, Vol. 111, No. 3 (January), pp. 621–688

Hills, Jill (2007): *Telecommunications and Empire* (Urbana-Champaign: University of Illinois Press)

Hoekman, Bernard and Aaditya Mattoo (2008): "Services Trade and Growth", in Juan A Marchetti and Martin Roy (eds.), *Opening Markets for Trade in Services: Countries and Sectors in Bilateral and WTO Negotiations* (Cambridge, UK: Cambridge University Press), pp. 21–58

International Telecommunication Union (ITU) (1984): *The Missing Link: Report of the Independent Commission for World-Wide Telecommunications Development* (Geneva: ITU)

Jain, R S (2001): "Spectrum Auctions in India: Lessons from Experience", *Telecommunications Policy*, Vol. 25, Nos. 10–11 (October–November), pp. 671–688

Joint Parliamentary Committee (JPC) to examine matters relating to Allocation and Pricing of Telecom Licences and Spectrum (JPC) (2013): *Report*, October 2013 (New Delhi: Lok Sabha Secretariat)

Kahn, Alfred E (1970): *The Economics of Regulation: Principles and Institutions, Volume I* (New York: John Wiley)

——— (1971): *The Economics of Regulation: Principles and Institutions, Volume II* (New York: John Wiley)

Kathuria, Rajat (2007): "Telecom Liberalization: A Case Study of India's Experience with Regulation", in Jaivir Singh (ed.), *Regulation, Institutions and the Law* (New Delhi: Social Science Press)

———, Mahesh Uppal, and Mamta (2009): "An Econometric Analysis of the Impact of Mobile", *India: The Impact of Mobile Phones, Vodaphone Policy Paper Series*, Number 9, pp. 5–20

Khilnani, Sunil (2009): "The Constitution and Individual Rights: A Comment on Dr. Abhishek Singhvi's *India's Constitution and Individual Rights: Diverse Perspectives*", *George Washington International Law Review*, Vol. 41, No. 2, pp. 361–366

Khurshid, Salman and Daksha Sharma (2018): *Spectrum Politics: Unveiling the Defence* (New Delhi: Rupa)

Klemperer, Paul (2002): "What Really Matters in Auction Design", *Journal of Economic Perspectives*, Vol. 16, No. 1 (Winter), pp. 169–189

Krishnaswamy, Sudhir (2013): "The Supreme Court on 2G: Signal and Noise", *Seminar*, Vol. 642, pp. 29–36

Kumar, Alok (2011): "3G Spectrum Auctions in India: A Critical Appraisal", *Economic and Political Weekly*, Vol. 46, No. 13 (March 26), pp. 121–129

Lessig, Lawrence (1999): "Commons and Code", *Fordham Intellectual Property, Media and Entertainment Law Journal*, Vol. 9, No. 2, pp. 405–419

______ (2006): *Code: Version 2.0* (New York: Basic Books)

Levi-Faur, David (1999): "The Governance of Competition: The Interplay of Technology, Economics, and Politics in European Union Electricity and Telecom Regimes", *Journal of Public Policy*, Vol. 19, No. 2, pp. 175–207

Levy, Brian and Pablo T Spiller (1994): "The Institutional Foundations of Regulatory Commitment: A Comparative Analysis of Telecommunications Regulation", *Journal of Law, Economics and Organization*, Vol. 10, No. 2 (October), pp. 201–246

Lok Sabha Debates (New Delhi: Lok Sabha Secretariat)

——— (LSD X 1995a): Tenth Series, Volume XLIII, No. 8, August 9, 1995, pp. 411–420

——— (LSD X 1995b): Tenth Series, Volume XLIII, No. 9, August 16, 1995, pp. 567–604

——— (LSD XI 1997): Eleventh Series, March 18, 1997

——— (LSD XIII 2000): Thirteenth Series, Volume V, No. 14, March 13, 2000, pp. 411–478

McCraw, Thomas K (1975): "Regulation in America: A Review Article", *Business History Review*, Vol. 49, No. 2 (Summer), pp. 159–183

McDowell, Stephen D and Jenghoon Lee (2003): "India's Experiments in Mobile Licensing", *Telecommunications Policy*, Vol. 27, Nos. 5–6 (June–July), pp. 371–382

Melody, William H (1997a): "Policy Objectives and Models of Regulation", in William H Melody (ed.), *Telecom Reform: Principles, Policies and Regulatory Practices* (Lyngby, Denmark: Den Private Ingeniørfond), pp. 11–24

——— (1997b): "Price Regulation and Its Implications", in William H Melody (ed.), *Telecom Reform: Principles, Policies and Regulatory Practices* (Lyngby, Denmark: Den Private Ingeniørfond), pp. 159–177

Mitnick, Barry M (1980): *The Political Economy of Regulation: Creating, Designing, and Removing Regulatory Forms* (New York: Columbia University Press)

Morgan, Bronwen (2003): "The Economization of Politics: Meta-Regulation as a Form of Nonjudicial Legality", *Social & Legal Studies*, Vol. 12, No. 4, pp. 489–523

——— and Karen Yeung (2007): *An Introduction to Law and Regulation* (Cambridge, UK: Cambridge University Press)

Mukherji, Rahul (2004): "Managing Competition: Politics and the Building of Independent Regulatory Institutions", *India Review*, Vol. 3, No. 4 (October), pp. 278–305

——— (2006): "Regulatory Evolution in Indian Telecommunications", *Institute of South Asian Studies Working Paper No. 7*, January [Available at

www.isas.nus.edu.sg/Attachments/PublisherAttachment/ISAS_working%20paper_7_21102009205358.pdf]
——— (2009): "Interests, Wireless Technology, and Institutional Change: From Government Monopoly to Regulated Competition in Indian Telecommunications", *Journal of Asian Studies*, Vol. 68, No. 2 (May), pp. 491–517
National Numbering Plan 2003 (NNP 2003): Department of Telecommunications, Government of India [Available at www.dot.gov.in]
National Telecom Policy 1994 (NTP 1994): Department of Telecommunications, Government of India [Available at www.dot.gov.in]
——— 1999 (NTP 1999): Department of Telecommunications, Government of India [Available at www.dot.gov.in]
——— 2012 (NTP 2012): Department of Telecommunications, Government of India [Available at www.dot.gov.in]
Neiman, Max (1980): "The Virtues of Heavy-Handedness in Government", *Law and Policy Quarterly*, Vol. 2, No. 1 (January), pp. 11–34
Noll, Roger (1983): "The Political Foundations of Regulatory Policy", *Journal of Institutional and Theoretical Economics*, Vol. 139, No. 3, pp. 377–404
______ and Scott J Wallsten (2006): "Universal Telecommunications Service in India", in Suman Bery, Barry Bosworth, and Arvind Panagariya (eds.), *India Policy Forum 2005–06, Volume 2* (New Delhi: Sage Publications), pp. 255–288
Ogus, Anthony (1995): "Rethinking Self-Regulation", *Oxford Journal of Legal Studies*, Vol. 15, No. 1, pp. 97–108
______ (2009): "Regulation Revisited", *Public Law*, Vol. 2, pp. 332–346
Orts, E W (1998): "Shirking and Sharking: A Legal Theory of the Firm", *Yale Law & Policy Review*, Vol. 16, No. 2, pp. 265–329
Parliamentary Debates: Rajya Sabha, Official Report (New Delhi: Rajya Sabha Secretariat)
——— (PRS 1995a): Volume CLXXIV, No. 12, August 22, 1995, pp. 353–368
——— (PRS 1995b): Volume 175, No. 2, 28 November, 1995, pp. 395–446
——— (PRS 1997): Volume 180, No. 21, 20 March, 1997, pp. 244–302
——— (PRS 2000): Volume 189, No. 16, 15 March, 2000, pp. 229–267
Peltzman, S (1998 [1989]): "The Economic Theory of Regulation after a Decade of Deregulation", in Robert Baldwin, Colin Scott, and Christopher Hood (eds.), *A Reader on Regulation* (Oxford: Oxford University Press), pp. 93–130
Petrazzini, Ben A (1996): "Telecommunications Policy in India: The Political Underpinnings of Reform", *Telecommunications Policy*, Vol. 20, No. 1 (January–February), pp. 39–51
Planning Commission, Government of India (PCGoI) (2008): *Approach to Regulation of Infrastructure* (New Delhi: Planning Commission, Government of India)
——— (2013): *Draft Regulatory Reform Bill* [Available at planningcommission.nic.in/reports/genrep/reg_bill2711.pdf [Accessed on December 15, 2013]

Posner, Richard A (1974): "Theories of Economic Regulation", *Bell Journal of Economics and Management Science*, Vol. 5, No. 2 (Autumn), pp. 335–358

Pound, Roscoe (1913): "Justice According to Law", *Columbia Law Review*, Vol. 13, No. 8 (December), pp. 696–713

——— (1914): "Justice According to Law", *Columbia Law Review*, Vol. 14, No. 2 (February), pp. 103–121

Prasad, Rohit (2011): "2008 Telecom Licensing Policy: Conceptual Issues", *Economic and Political Weekly*, Vol. 46, No. 53 (December 31), pp. 23–26

——— and V Sridhar (2014): *The Dynamics of Spectrum Management: Legacy, Technology, and Economics* (New Delhi: Oxford University Press)

Prosser, Tony (1999): "Theorising Utility Regulation", *Modern Law Review*, Vol. 62, No. 2 (March), pp. 196–217

______ (2006): "Regulation and Social Solidarity", *Journal of Law and Society*, Vol. 33, No. 3, pp. 364–387

Raghavan, Vikram (2007): *Communications Law in India (Legal Aspects of Telecom, Broadcasting, and Cable Services)* (New Delhi: LexisNexis Butterworths)

——— (2012): "Beyond the Headlines of the 2G Reference Opinions", *Economic and Political Weekly, Web Exclusives*, Vol. 47, No. 43 (October 27) [Available at www.epw.in]

Renda, Andrea (2012): "Telecommunications Regulation", in Roger J Van den Bergh and Alessio M Pacces (eds.), *Regulation and Economics (Encyclopedia of Law and Economics, Second Edition, Volume 9)* (Cheltenham: Edward Elgar)

Röller, Lars-Hendrik, and L Waverman (2001): "Telecommunications Infrastructure and Economic Development: A Simultaneous Approach", *American Economic Review*, Vol. 91 No. 4 (September), pp. 909–923

Roseman, Daniel (2003): "Domestic Regulation and Trade in Telecommunications Services: Experience and Prospects under the GATS", in Aditya Mattoo and Pierre Sauvé (eds.), *Domestic Regulation & Service Trade Liberalization* (Washington, DC: The World Bank)

Rossi, Jim (1997): "Participation Run Amok: Thee Costs of Mass Participation for Deliberative Agency Decisionmaking", *Northwestern University Law Review*, Vol. 92, No. 1, pp. 173–249

Ryan, Patrick S (2004): "Application of the Public-Trust Doctrine and Principles of Natural Resource Management to Electromagnetic Spectrum", *Michigan Telecommunications and Technology Law Review*, Vol. 10, No. 2 (Spring), pp. 285–372

Saunders, Robert J, J J Warford, and B Wellenius (1983): *Telecommunications and Economic Development* (World Bank, Baltimore: The Johns Hopkins University Press)

Sax, Joseph L (1970): "The Public Trust Doctrine in Natural Resource Law: Effective Judicial Intervention", *Michigan Law Review*, Vol. 68, pp. 471–566

Selznick, Philip (1985): "Focusing Organizational Research on Regulation", in Roger G Noll (ed.), *Regulatory Policy and the Social Sciences* (Berkeley: University of California Press), pp. 363–368

Shepsle, Kenneth A (1989): "Studying Institutions: Some Lessons from the Rational Choice Approach", *Journal of Theoretical Politics*, Vol. 1, No. 2 (April), pp. 131–147

Singh, J P (1998): "Unravelling 'the missing link': The Provision of Telecommunication Services in Select Developing Countries", *Communicatio*, Vol. 24, No. 1, pp. 48–58

——— (2000): "The Institutional Environment and Effects of Telecommunication Privatization and Market Liberalization in Asia", *Telecommunications Policy*, Vol. 24, Nos. 10–11 (November), pp. 885–906

Singh, Jaivir (ed.) (2007): *Regulation, Institutions and the Law* (New Delhi: Social Science Press)

Singh, Rahul (2009): "The Teeter-Totter of Regulation and Competition: Balancing the Indian Competition Commission with Sectoral Regulators", *Washington University Global Studies Law Review*, Vol. 8, No. 7, pp. 71–105

Sinha, Nikhil (1996): "The Political Economy of India's Telecommunication Reforms", *Telecommunications Policy*, Vol. 20, No. 1 (January–February), pp. 23–38

Skocpol, Theda (1985): "Bringing the State Back in: Strategies of Analysis in Current Research", in Peter B Evans, Dietrich Rueschemeyer, and Theda Skocpol (eds.), *Bringing the State Back in* (Cambridge, UK: Cambridge University Press), pp. 3–37

Sørensen, Eva and Jacob Torfing (2004): "Making Governance Networks Democratic", *Working Paper 2004: 1* (Roskilde, Denmark: Centre for Democratic Network Governance, Roskilde University)

Sridhar, Kala Seetharam and V Sridhar (2004): "Telecommunications Infrastructure and Economic Growth: Evidence from Developing Countries", *Working Paper No. 14* (New Delhi: National Institute of Public Finance and Policy)

Sridhar, Varadharajan (2012): *The Telecom Revolution in India: Technology, Regulation, and Policy* (New Delhi: Oxford University Press)

——— and Rohit Prasad (2011): "Towards a New Policy Framework for Spectrum Management in India", *Telecommunications Policy*, Vol. 35, No. 2 (March), pp. 172–184

Standing Committee on Communications, 1996–97 (SCC 96–97) (1996): *Fourth Report: Telecom Regulatory Authority of India Bill, 1996* (New Delhi: Lok Sabha Secretariat)

———, 1999–2000 (SCC 99–00) (2000a): *Fifth Report: Action Taken by the Government on the Recommendations/ Observations Contained in the Sixth Report (Twelfth Lok Sabha) of Standing Committee on Communications on Recommendations of Telecom Regulatory Authority of India (TRAI) Regarding Increase in Telephone Tariff of Basic Telecom Services* (New Delhi: Lok Sabha Secretariat)

——— (2000b): *Tenth Report: Demands for Grants (2000–2001)* (New Delhi: Lok Sabha Secretariat)

Standing Committee on Information Technology, 1999–2000 (SCIT 99–00) (2000): *Thirteenth Report: Functioning of Telecom Regulatory Authority of India* (New Delhi: Lok Sabha Secretariat)

——— (2001a): *Limited Mobility Through WLL for Fixed Service Providers* (New Delhi: Lok Sabha Secretariat)

——— (2001b): *Implementation of New Telecom Policy, 1999* (New Delhi: Lok Sabha Secretariat) [Available at 164.100.47.133/ls/committeeR/Communication/27.pdf]

——— (2001c): *The Communication Convergence Bill, 2001: REPORT* (New Delhi: Lok Sabha Secretariat)

———, 2003 (SCIT 03) (2003): *Fifty-Third Report: Action Taken by Government on the Recommendations/Observations Contained in the Thirteenth Report (Thirteenth Lok Sabha) of the Standing Committee on Information Technology on "Functioning of Telecom Regulatory Authority of India (TRAI)"* (New Delhi: Lok Sabha Secretariat)

———, 2006–7 (SCIT 06–07) (2006a): *Thirty-Fifth Report: Action Taken by Government on the Recommendations/ Observations of the Committee Contained in Their Seventeenth Report (Thirteenth Lok Sabha) on 'Limited Mobility Through WLL for Fixed Service Providers'* (New Delhi: Lok Sabha Secretariat)

——— (2006b): *Thirty-Sixth Report: Action Taken by Government on the Recommendations/ Observations of the Committee Contained in Their Twenty-Eighth Report (Fourteenth Lok Sabha) on 'Spectrum Management'* (New Delhi: Lok Sabha Secretariat)

———, 2008–9 (SCIT 08–09) (2009): *Sixty-Eighth Report: The Telecom Regulatory Authority of India (Amendment) Bill, 2008* (New Delhi: Lok Sabha Secretariat)

Stason, E Blythe (1938): "Administrative Tribunals: Organization and Reorganization", *Michigan Law Review*, Vol. 36, No. 4 (February), pp. 533–567

Stewart, Richard B (1975): "The Reformation of American Administrative Law", *Harvard Law Review*, Vol. 88, No. 8 (June), pp. 1667–1813

Stigler, George J (1971): "The Theory of Economic Regulation", *Bell Journal of Economics and Management Science*, Vol. 2, No. 1 (Spring), pp. 3–21

_______ and Claire Friedland (1962): "What Can Regulators Regulate? The Case of Electricity", *Journal of Law and Economics*, Vol. 5 (October), pp. 1–16

Stone Sweet, Alec (1999): "Judicialization and the Construction of Governance", *Comparative Political Studies*, Vol. 32, No. 2 (April), pp. 147–184

Strange, Susan (1996): *The Retreat of the State: The Diffusion of Power in the World Economy* (Cambridge, UK: Cambridge University Press)

Sunstein, Cass R (1990): *After the Rights Revolution: Reconceiving the Regulatory State* (Cambridge, MA: Harvard University Press)

Swaminathan, Rajesh (1997): " 'Functioning Anarchy': India's National Telecommunications Policy and the Development of 'Basic' Telephone Services", *Columbia Journal of Asian Law*, Vol. 11, No. 2, pp. 393–433

Tangirala, Maruthi P (2009): "Development Impact of Mobile Telephony", M Phil Dissertation, Indian Institute of Public Administration, New Delhi/Panjab University, Chandigarh (unpublished)

——— (2015): "The Costs of Inappropriate Delegation: TRAI Act and Its Amendment", *Competition and Regulation in Network Industries*, Vol. 16, No. 1, pp. 47–65

——— (2016): "Call Drops and Regulatory Costs", *Economic and Political Weekly*, Vol. 51, No. 13, pp. 19–20

Telecom Regulatory Authority of India (TRAI) (1998): *Orders: 1997–98* (New Delhi: TRAI)

——— (1999a): *Annual Report: 1997–98* (New Delhi: TRAI)

——— (1999b): *Telecommunication Tariff Order 1999*, March 9, 1999 [Available at www.trai.gov.in]

——— (1999c): *Telecommunication Interconnection (Charges and Revenue Sharing) Regulation 1999*, May 28, 1999 [Available at www.trai.gov.in]

——— (1999d): *Consultation Paper on MTNL's Tariff Proposal for Cellular Mobile Services Using C.D.M.A. Technology and Related Issues*, October 12, 1999 [Available at www.trai.gov.in]

——— (1999e): *Recommendations on License Fee and Terms and Conditions of the License Agreement for GMPCS Service*, November 15, 1999 [Available at www.trai.gov.in]

——— (2000a): *Recommendations on Issues Relating to Cellular Mobile Service*, June 23, 2000 [Available at www.trai.gov.in]

——— (2000b): *Annual Report: 1998–99* (New Delhi: TRAI)

——— (2000c): *Recommendations on Issues Relating to Licensing of Fixed Service Providers*, August 31, 2000 [Available at www.trai.gov.in]

——— (2000d): *TRAI Response on Basic Services Recommendations Referred by DoT for Reconsideration*, October 31, 2000 [Available at www.trai.gov.in]

——— (2000e): *Consultation Paper on Policy Issues Relating to Limited Mobility by Use of Wireless in Local Loop Techniques in the Access Network by Basic Service Providers*, No. 2005/5-FN, November 3, 2000 [Available at www.trai.gov.in]

——— (2001a): *Recommendations on Issues Relating to "Limited Mobility" Through Wireless in Local Loop in the Access Network by Basic Service Providers*, January 8, 2001, pp. 74–105 [Available on file with author]

——— (2001b): *Recommendations of the TRAI on Universal Service Obligations*, October 3, 2001 [Available at www.trai.gov.in]

——— (2003a): *Telecommunication Interconnection Usage Charges (IUC) Regulation 2003*, January 24, 2003 [Available at www.trai.gov.in]

——— (2003b): *Consultation Paper on Unified Licensing for Basic and Cellular Services*, July 16, 2003 [Available at www.trai.gov.in]

——— (2003c): *Recommendations on Unified Licensing*, October 27, 2003 [Available at www.trai.gov.in]

——— (2003d): *Telecommunication Interconnection Usage Charges Regulation 2003*, October 29, 2003 [Available at www.trai.gov.in]

——— (2004a): *The Indian Telecom Services Performance Indicators Oct–Dec '03*, March, 2004 [Available at www.trai.gov.in

——— (2004b): *Consultation Paper on Spectrum Related Issues: Efficient Utilisation, Spectrum Allocation, and Spectrum Pricing*, May 31, 2004 [Available at www.trai.gov.in]

——— (2004c): *The Indian Telecom Service Performance Indicators Jan-Mar '04*, June 2004 [Available at www.trai.gov.in]

——— (2005a): *Recommendations on Unified Licensing*, January 13, 2005 [Available at www.trai.gov.in]

——— (2005b): *The Indian Telecom Services Performance Indicators Oct-Dec '04*, March 2005 [Available at www.trai.gov.in]

——— (2005c): *Recommendations on Spectrum Related Issues*, May 13, 2005 [Available at www.trai.gov.in]

——— (2005d): *Study Paper on 'Indicators for Telecom Growth'*, No. 2/2005 [Available at www.trai.gov.in/trai/upload/studypapers/2/ir30jun.pdf]

——— (2006a): *The Indian Telecom Services Performance Indicators October-December 2005*, April 10, 2006 [Available at www.trai.gov.in]

——— (2006b): *Recommendations on Components of Adjusted Gross Revenue (AGR)*, September 13, 2006 [Available at www.trai.gov.in]

——— (2006c): *Recommendations on Allocation and Pricing of Spectrum for 3G and Broadband Wireless Access Services*, September 27, 2006 [Available at www.trai.gov.in]

——— (2007a): *The Indian Telecom Service Performance Indicators October-December 2006*, April 17, 2007 [Available at www.trai.gov.in]

——— (2007b): *Consultation Paper on Review of License Terms and Conditions and Capping of Number of Access Providers*, June 12, 2007 [Available at www.trai.gov.in]

——— (2007c): *Recommendations on Review of License Terms and Conditions and Capping of Number of Access Providers*, August 28, 2007 [Available at www.trai.gov.in]

——— (2007d): *A Journey Towards Excellence in Telecommunications*, September 21, 2007 [Available at www.trai.gov.in]

——— (2007e): *The Indian Telecom Services Performance Indicators April-June 2007*, October 5, 2007 [Available at www.trai.gov.in]

——— (2008a): *Recommendations on Permitting New Entity for Allocation of 3G Spectrum*, April 25, 2008 [Available at www.trai.gov.]

——— (2008b): *The Indian Telecom Services Performance Indicators October–December 2007*, April, 2008 [Available at www.trai.gov.in]

——— (2008c): *Consultation Paper on Allocation and Pricing for 2.3–2.4 GHZ, 2.5–2.69 GHZ & 3.3–3.6 GHZ Bands*, May 2, 2008 [Available at www.trai.gov.in]

——— (2008d): *Recommendations on Allocation and Pricing for 2.3–2.4 GHz, 2.5–2.69 GHz & 3.3–3.6 GHz Bands*, July 11, 2008 [Available at www.trai.gov.in]

——— (2008e): *Recommendations on Modifications Proposed by DoT with Regard to 3G Services, DO No. 103–5/2008-MN*, July 12, 2008 [Available at www.trai.gov.in]

——— (2008f): *Recommendations on Modifications Proposed by DoT with Regard to SUC, DO No. 101–19/2007-MN (Vol. III)*, July 16, 2008 [Available at www.trai.gov.in]

——— (2008g): *Recommendations on the Reference on the Spectrum Charging for 3G Services*, December 9, 2008 [Available at www.trai.gov.in]

——— (2008h): *The Telecommunications Unsolicited Commercial Communications (Amendment) Regulations 2008*, March 17, 2008 [Available at www.trai.gov.in]

——— (2009): *Comments on the Reference on the Auctioning Procedure and Reserve Price of One Carrier 2 x 1.25 MHz for EVDO in 800 MHz Band for CDMA 3G Services*, January 23, 2009 [Available at www.trai.gov.in]

——— (2010): *Recommendations on Spectrum Management and Licensing Framework*, May 11, 2010 [Available at www.trai.gov.in]

——— (2011a): "Telecom Subscription Data as on 30th November, 2010", *Press Release no. 5/2011*, January 25, 2011 [Available at www.trai.gov.in]

——— (2011b): *Letter no. 103–2/2011-MN Dated 8th February, 2011* [Available at www.trai.gov.in]

——— (2011c): *Report on the 2010 Value of Spectrum in 1800 MHz Band*, January 30, 2011 [Available at www.trai.gov.in]

——— (2012a): "Highlights on Telecom Subscription Data as on 31st March, 2012", *Press Release*, May 3, 2012 [Available at www.trai.gov.in]

——— (2012b): "Monthly Highlights on Telecom Subscription Data as on 31st July 2012", *Press Release No. 179/2012*, September 7, 2012 [Available at www.trai.gov.in]

——— (2012c): *Telecom Sector in India: A Decadal Profile* (New Delhi: TRAI)

——— (2012d): *Annual Report: 2011–12* (New Delhi: TRAI)

——— (2012e): *The Telecommunication Tariff (Fifty Second Amendment) Order 2012*, September 19, 2012 [Available at www.trai.gov.in]

——— (2012f): *Telecommunications Mobile Number Portability (4th Amendment) Regulations 2012*, September 19, 2012 [Available at www.trai.gov.in]

——— (2012g): *The Reporting System on Accounting Separation (Amendment) Regulations 2012*, October 15, 2012 [Available at www.trai.gov.in]

——— (2012h): *The Standards of Quality of Service of Basic Telephone Service (Wireline) and Cellular Mobile Telephone Service (Second Amendment) Regulations 2012*, November 8, 2012 [Available at www.trai.gov.in]

——— (2013a): *Consultation Paper on Valuation and Reserve Price of Spectrum*, July 23, 2013 [Available at www.trai.gov.in]

——— (2013b): *Recommendations on Valuation and Reserve Price of Spectrum*, September 30, 2013 [Available at www.trai.gov.in]

——— (2015a): *Recommendations on Definition of Revenue Base (AGR) for the Reckoning of Licence Fee and Spectrum Usage Charges*, January 6, 2015 [Available at www.trai.gov.in]

——— (2015b): *Highlights of Telecom Subscription Data as on 31st March, 2015*, May 12, 2015 [Available at www.trai.gov.in]

Thakurta, Paranjoy Guha and Aditi Roy Ghatak (2012): "TRAI Recommendations: The Next Round of Telecom Tussles", *Economic and Political Weekly*, Vol. 47, No. 22, pp. 51–57

Thatcher, Mark and Alec Stone Sweet (2002): "Theory and Practice of Delegation to Non-Majoritarian Institutions", *West European Politics*, Vol. 25, No. 1 (January), pp. 1–22

Thiruvengadam, Arun K and Piyush Joshi (2012): "Judiciaries as Crucial Actors in Southern Regulatory Systems: A Case Study of Indian Telecom Regulation", *Regulation & Governance*, Vol. 6, No. 3 (September), pp. 327–343

Tuthill, Lee (1997): "The GATS and New Rules for Regulators", *Telecommunications Policy*, Vol. 21, Nos. 9–10 (November–December), pp. 783–798

——— and Laura B Sherman (2008): "Telecommunications: Can Trade Agreements Keep Up with Technology?" in Juan A Marchetti and Martin Roy (eds.), *Opening Markets for Trade in Services: Countries and Sectors in Bilateral and WTO Negotiations* (Cambridge, UK: Cambridge University Press), pp. 151–183

Uppal, Mahesh, S K N Nair, and C S Rao (2006): *India's Telecom Reform: A Chronological Account* (New Delhi: UNDP/IIPA)

Vakil, Raeesa (2013): "Advising the President", *Seminar*, Vol. 642 (February) [Available at www.india-seminar.com/cd8899/cd_frame8899.html]

Veljanovski, Cento (2010): "Economic Approaches to Regulation" in Robert Baldwin, Martin Cave, and Martin Lodge (eds.), *The Oxford Handbook of Regulation* (Oxford: Oxford University Press), pp. 17–38

Venugopal, Krishnan (2003): "Telecommunication Sector Negotiations at the WTO: Case Studies of Indian, Sri Lanka and Malaysia", *ITU/ESCAP/WTO Regional Seminar on Telecommunications and Trade Issues*, 28–30 October 2003, Bangkok, Thailand [Available at www.unescap.org/tid/mtg/ituwtoesc_s51b.pdf]

Walden, Ian and John Angel (2007): *Telecommunications Law and Regulation* (New Delhi: Oxford University Press)

Waverman, Leonid, M Meschi, and M Fuss (2005): "The Impact of Telecoms on Economic Growth in Developing Countries", *Africa: The Impact of Mobile Phones, Vodaphone Policy Paper Series*, No. 3 (March), pp. 10–23

Weingast, Barry R (1981): "Regulation, Reregulation, and Deregulation: The Political Foundations of Agency Clientele Relationships", *Law and Contemporary Problems*, Vol. 44, No. 1 (Winter), pp. 147–177

Wellenius, Björn and Peter A Stern (1994): "Implementing Reforms in the Telecommunications Sector: Background, Overview, and Lessons", in *Implementing Reforms in the Telecommunications Sector: Lessons from Experience* (Washington, DC: World Bank), pp. 1–63

Werbach, Kevin (2004): "Supercommons: Towards a Unified Theory of Wireless Communication", *Texas Law Review*, Vol. 82, pp. 863–973

——— (2011): "The Wasteland: Anticommons, White Spaces, and the Fallacy of Spectrum", *Arizona Law Review*, Vol. 53, No. 1, pp. 213–254

Williamson, Oliver E (1985): *The Economic Institutions of Capitalism: Firms, Markets, Relational Contracting* (New York: The Free Press)

——— (1996): *The Mechanisms of Governance* (New York: Oxford University Press)

——— (2000): "The New Institutional Economics: Taking Stock, Looking Ahead", *Journal of Economic Literature*, Vol. 38, No. 3 (September), pp. 595–613

Court cases

Associated Provincial Picture Houses v. *Wednesbury Corp.* (1947) 2 All ER 680

Attorney General v. *Edison Telephone Co of London*, [1880]6 QBD 244

Bharti Cellular Limited v. *Union of India and Ors.* (2010) 10 SCC 174

Brahm Dutt v. *Union of India*, AIR 2005 SC 730

Cellular Operators Association of India and Ors. v. *Union of India and Ors.* (2003) 3 SCC 186

Centre for Public Interest Litigation v. *Union of India and Ors.* (2012) 3 SCC 1

Clariant International Ltd. and Anr. v. *Securities and Exchange Board of India*, AIR 2004 SC 4236

Competition Commission of India v. *Steel Authority of India and Anr.*, 2010 CompLR 0061 (SC)

Delhi Science Forum and Ors. v. *Union of India and Anr.*, AIR 1996 SC 1356

Dr Subramanian Swamy v. *Dr Manmohan Singh and Anr.* (2012) 3 SCC 64

Fomento Resorts and Hotels Limited v. *Minguel Martins* (2009) 3 SCC 571

Illinois Central Railroad Co. v. *People of the State of Illinois*, 146 US 387 (1892)

In re: Special Reference No. 1 of 2012 (2012) 10 SCC 1

M C Mehta v. *Kamal Nath* (1997) 1 SCC 388

MCI Communications Corp. v. *AT&T Co.*, 708 F.2d 1081, 1132–33 (7th Cir. 1983)

Microwave Communication Limited v. *Union of India and Ors.*, 1999 Indlaw Del 11471

MTNL v. *TRAI* and *Telecom Watchdog* v. *TRAI*, 2000 Indlaw Del 330

Munn v. *Illinois*, 94 US 113 (1877)

Shyam Telelink Limited v. *Union of India* (2010) 10 SCC 165

Tata Cellular v. *Union of India*, AIR 1996 SC 11

Union of India v. *Telecom Regulatory Authority of India*, 1998 Indlaw Del 562

Union of India and Anr. v. *Association of Unified Telecom Service Providers of India and Ors.* (2011) 10 SCC 543
United States v. *AT&T*, 552 F. Supp. 131 (1982)
Verizon Communications Inc. v. *Law Offices of Curtis V. Trinko LLP*, 540 US 398 (2004)
West Bengal Electricity Regulatory Commission v. *CESC Ltd.*, AIR 2002 SC 3588

INDEX

For Product Safety Concerns and Information please contact our EU representative GPSR@taylorandfrancis.com Taylor & Francis Verlag GmbH, Kaufingerstraße 24, 80331 München, Germany

Batch number: 10725981

Printed by Printforce, the Netherlands